200 Years American Manufactured Jewelry & Accessories

Suzanne Marshall

with A Special Chapter on Button History
by Eric W. Reichardt

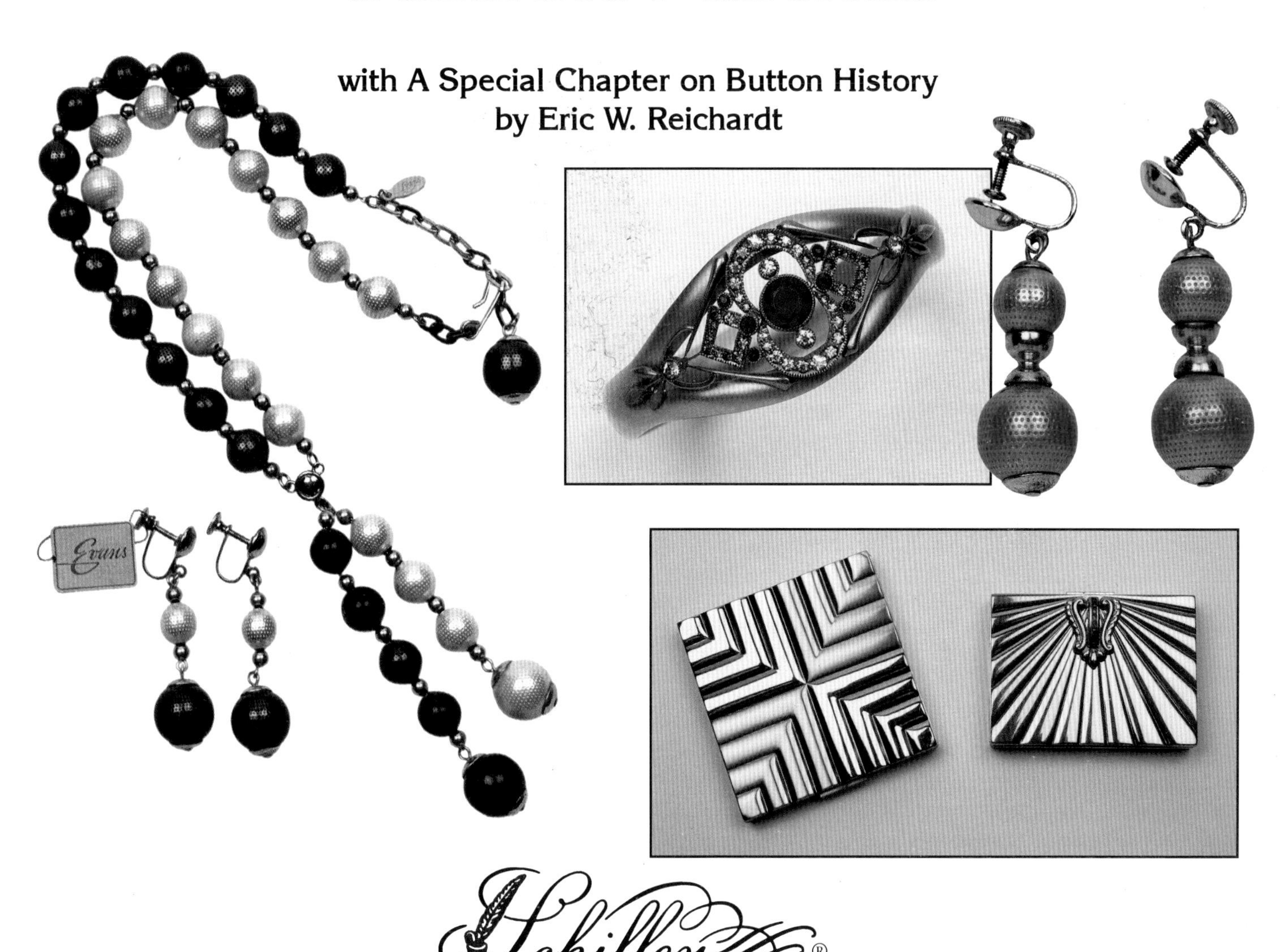

Schiffer Publishing Ltd ®

4880 Lower Valley Road, Atglen, PA 19310 USA

Library of Congress Cataloging-in-Publication Data

Marshall, Suzanne.
200 years of American manufactured jewelry & accessories / by Suzanne Marshall.
p. cm.
ISBN 0-7643-1838-1 (pbk.)
1. Jewelry--Collectors and collecting--United States--Catalogs. 2. Dress accessories--Collectors and collecting--United States--Catalogs. I. Title: Two hundred years of American manufactured jewelry & accessories. II. Title.
NK7312.M37 2003
688'.2'0973--dc21

2003007377

Designed by Mark David Bowyer
Type set in Korinna BT / Korinna BT

ISBN: 0-7643-1838-1
Printed in China
1 2 3 4

Published by Schiffer Publishing Ltd.
4880 Lower Valley Road
Atglen, PA 19310
Phone: (610) 593-1777; Fax: (610) 593-2002
E-mail: Info@schifferbooks.com
Please visit our web site catalog at
www.schifferbooks.com
We are always looking for people to write books on new and related subjects. If you have an idea for a book please contact us at the above address.

This book may be purchased from the publisher.
Include $3.95 for shipping.
Please try your bookstore first.
You may write for a free catalog.

In Europe, Schiffer books are distributed by
Bushwood Books
6 Marksbury Ave.
Kew Gardens
Surrey TW9 4JF England
Phone: 44 (0) 20 8392-8585; Fax: 44 (0) 20 8392-9876
E-mail: Bushwd@aol.com
Free postage in the U.K., Europe; air mail at cost.

Contents

Acknowledgments

The author appreciates the generous assistance of many people who have graciously loaned items from their collections, provided historical information, and assisted in numerous ways for the completion of this book:

Thank you to George Sparacio, Eric W. Reichardt, Richard Waks, Laurie-Ann Ackerman-Marsh Ackerman, Ltd., Patricia Berry, Christine L. Rogers, Merle S. Koblenz, Paula Deane, Bruce Cherner Antique Silver, Jacquelyn-Naples, Florida, Lalla Rookh Boutique-Naples, Florida, and Andy R, all of whom have loaned items from their collections for photography. Some collectors have chosen to remain anonymous-their jewelry is equally appreciated.

Also a special thank you to Robert E. Mower, Edward Stevenson, Robert W. Lanpher, George Cunningham and Nancy Campbell of the North Attleboro Falls Museum, Elizabeth Korostynski, Joseph J. MacDougald, Joseph Viveiros, Christopher Sweet, Mel Allenson of Swank, Inc, and Dorothy Turini for providing historical information, early catalogs and trade journals and for assisting in numerous ways.

Specific photograph credits are for the Plainville Stock Company filigree necklace, courtesy of Terry Sue Tyrrell Guedri, photography by Todd Torgersen; Paye and Baker Billiken pin, courtesy of Christie Romero, *Warman's Jewelry 3rd edition,* Krause Publications, 2002; and Paye and Baker advertisements from *Silver Magazine,* September/October 1990 issue, reprinted with permission of *Silver Magazine.* Some other photographs were taken by Bruce Waters of Schiffer Publishing.

The Joseph L. Sweet Memorial Library and the Attleboro Area Industrial Museum of Attleboro, Massachusetts have provided historical information. My parents, Patricia and William Berry, have assisted with research. My brother Lincoln B. Berry was the photographer for the majority of items in this book. Thos. Moser provided artwork. Richard McMullen contributed enamel history. A special thank you to Douglas Congdon-Martin of Schiffer Publishing. My long-time friend and mentor, Gary Beckwith, has taught me the processes of jewelry making and offered encouragement. Other friends have helped in numerous ways-thank you to each of you.

Chapter 1

Attleboro and Its Place in American Jewelry Manufacturing History

The original town of Attleborough, named after its sister city in England, was a large area including what is now Attleboro, North Attleboro, and Rehoboth, Massachusetts. Pilgrims from the Plymouth colonies settled this agricultural area in the mid-1600s, joining an American Indian population with a rich heritage.

Abundant natural resources, including rich soil and beautiful streams and falls, helped to shape the destiny of our forefathers and their lives in this early American settlement. Farming and textiles were early occupations.

The history of jewelry crafting began around 1780, when an unknown Frenchman, said to have fled Lafayette's Continental Army, built a small shop near the corner of South Washington and Chestnut Streets in what is now North Attleboro. History refers to this early jewelry maker as "the Frenchman," and he is still honored by a plaque near the site of this first jewelry shop.

Around 1793, Edward Price traveled from England and set up a shop to make buttons. The original Price home is still standing in North Attleboro. American buttons from this period are unmarked by their makers; therefore we can only offer examples of the type of button, which would have been made by Edward Price.

Buttons were a very successful form of metal manufacturing in Attleboro. The early 1800s brought the firms of Robinson Jones and D. Evans, who manufactured military, uniform, and fancy civilian buttons with a wonderful gold finish. In the 1830s, the early days of America's industrial revolution, national awards from the American Institute celebrated the high quality of these buttons with commemorative medals.

The first half of the nineteenth century were formative years for jewelry manufacturing in Attleboro, as well as in Providence, Rhode Island. The Boston and Providence Highway, known today as Route 1, was an early roadway connecting the cities and running directly through North Attleboro.

Jewelry manufacturing was a "grass roots" venture. With high hopes for success, some individuals set up shop in their homes, sheds, or barns, while others formed partnerships and built factories. Secrets of gold electroplating were shared between local jewelry makers, with the latest techniques brought back to Attleboro by young men traveling the twelve miles south to Providence, a more metropolitan community.

Jewelry for women in the 18th and early 19th century may have included a simple gold wedding band, beads of gold or coral, or a mourning brooch. During the industrial revolution and throughout the nineteenth century in America, women began collecting jewelry for fashion. Attleboro makers perfected 10K and 14K gold, gold electroplate, and gold-filled manufacturing techniques in the 1800s. With the talents of jewelry designers, tool and die makers, finishers, engravers, and others involved in jewelry manufacture from start to finish, the Attleboro companies experienced great success. Some opened offices in New York, Chicago, and San Francisco, and many shipped their jewelry worldwide.

During the Victorian era, jewelry manufacturing using the gold electroplate and gold-filled processes made jewelry far more available to the average, middle class American consumer. Elegantly beautiful designs, in what is now called "Victorian costume jewelry," were plentiful and affordable. Women began collecting jewelry after the Civil War and especially after 1870. Colorful fashion prints, as seen in Godey's Ladies Book, were an inspiration of the Victorian style in clothing and jewelry.

During most of the nineteenth century, Attleboro manufacturers generally did not mark their jewelry with company trademarks. For this reason, there is a relatively small amount of early jewelry that can be documented, although we know from old histories and trade journals that large quantities of jewelry were produced. One of the early companies to mark their wares is the Plainville Stock Company, which advertised their trademark, "P.S.CO," in an 1886 trade journal. Included in this book is one of their brooches in the Etruscan Revival style, as well as a black onyx cameo scarf pin. Another old firm, Smith and Crosby placed their "S & C" trademark and a patent date of 1880 on a pair of gold engraved cufflinks with adjustable backs, marked and dated for patent purposes. Trademarks became prevalent around 1890 on Attleboro jewelry and by the early 1900s appear to have been widely used.

Around 1890 sterling silver experienced a great rise in popularity. Jewelry, souvenir spoons, match safes, luggage tags, thimbles and other sewing items, napkin rings, pencil cases, letter openers and a large selection of silver novelties from this period have become very collectible today. Some Attleboro makers were already prominent in the manufacture of sterling silver flatware and hollowware, and were able to expand their production into other areas during this time.

The 1890s were also a popular period for novelty jewelry with enameling. The Palmer Cox Brownies were popular and colorful characters, and the Regnell and Bigney

Company advertised enameled versions of them in 1894, during Attleboro's 200th anniversary. Enamel flag pins and badges for schools, with a choice of "any letters or figures of raised metal," were advertised as a "roaring success" in the 1897 McRae & Keeler catalog.

McRae & Keeler's 64-page catalog of 1897 states on the opening page "Situated in the largest jewelry centre of this country (if not in the world) employing the finest workmen in every branch of the trade, using the latest improved machinery and producing goods at the lowest minimum figure possible. We are the only actual manufacturers and practical jewelers issuing a catalog." The catalog shows solid gold rings and patterned wedding bands and a wide assortment of watch chains in gold filled, some with intaglio fobs. Pocket watches were illustrated with gold-filled cases by Bates and Bacon, another Attleboro manufacturer. Cufflinks, scarf pins, shirt studs and waist sets, emblem buttons, society pins and badges, belt buckles, and flags for all nations are examples of this catalog's diversity. In this 1897 catalog the author discovered the "Black Cat Stickpin," enamel on gold plate with "emerald eyes," and realized the same stickpin was in her grandmother's jewelry box.

Sterling silver Art Nouveau match safes, designed to keep wooden stick matches dry and to prevent them from igniting in the wearer's pocket, were at the height of their usefulness and fashion, between 1893 and about 1910. Numerous Attleboro makers offered beautifully designed match safes, often with figural scenes reminiscent of the time. The firm of F.M. Whiting & Co. made a match safe with a cigar cutter attachment. R. Blackinton & Co. issued a catalog of "Match Boxes—Made in Sterling Silver" c. 1905, showing 110 of their match safes, priced from $1.63-5.00. They also offered five of these match safes in 14K gold for $22.50-28.50. The Paye & Baker catalog of 1908 advertises match boxes with "good weight, excellent workmanship and of unique design" and most "curved to fit the pocket." Although sterling silver was the metal of choice for match safes, two Attleboro firms patented silver alloys for their manufacture and called them "Sterline" and "Silveroin."

James E. Blake patented the alloy "Sterline" for the manufacture of Art Nouveau jewelry, match safes, and dresser ware, c. 1900. Sterline is a non-tarnishing silver metal resembling sterling silver. The Bristol Silver Company offered their own silver alloy resembling sterling silver and called it "Silveroin," the mark being seen on "toilet ware and novelties" as stated by the Jewelers Circular catalog of 1915. Their line of mesh purses is marked "German Silver," an alloy of nickel, copper, and zinc, which in combination has a silver appearance.

Many examples of New England ingenuity are found in Attleboro jewelry history. The J.F. Sturdy Company built a machine for the manufacture of "ladies eyeglass chains" around 1900 at the height of the Gibson Girl's popularity, a beautiful woman with an upswept hairdo, whose image was created in the drawings of Charles Dana Gibson. The fashion was to wear a gold hairpin with a delicate attached chain, which joined the lady's gold framed spectacles. When styles changed and eyeglass chains were no longer popular, Sturdy converted its eyeglass chain machine to manufacture collar pins for men, with excellent success.

Sturdy Factory, Attleboro Falls, c. 1910. This firm manufactured gold-filled jewelry, including chains, lockets, and StaLokt cufflinks.

Around 1900, there were 53 operating jewelry manufacturers in the Attleboro area. North Attleboro and Attleboro had become two separate towns in 1887, and the latter incorporated as a city in 1914 and its name was officially changed to Attleboro. "The Attleboro's" are generally referred to with the shortened spelling today, and the spelling "Attleboro" will be used in this book for simplicity, when referring to the city of Attleboro and to both Attleboro Falls and North Attleboro.

Attleboro was advertised as "The Hub of the Jewelry World" on a sign at the Attleboro Depot about 1918, when jewelry manufacturing was by far the predominant industry in the area. The Edwardian gentleman or lady was offered a very wide selection of watch chains, bangle bracelets, sash pins, lace pins, scarf pins, cufflinks, and cloisonné enamel jewelry.

The jewelry industry was considered an excellent place of employment with better pay and shorter hours than other industries of the time. Many people traveled from Europe and England to settle in Attleboro in the late 1800s and to work in the jewelry factories. Some craftsmen who started at the jeweler's bench later went on to own their own companies. Attleboro jewelry firms were family owned businesses, creating stability for employment.

Groups of women commuted each weekday from the city of Taunton to work in the Attleboro jewelry factories. Some jewelry makers offered four-year apprenticeships.

Enamel jewelry was prominent from the 1890s onward. There was always competition in the enamel departments to do the best work, as salaries were based on an individual's talent, therefore producing very high quality enamel items. Women were especially talented in enamel, which combined the manufacturing process with hand work in Attleboro. For enamels, the jewelry was die struck, often with a decorative pattern stamped into the metal. The enamel was applied by hand, and could require as many as twenty firings in a kiln to achieve the many colors and effects on the finished "work of art." Translucent enamels were applied over the stamped patterns to create *guilloche*, a technique frequently seen in Attleboro enamel jewelry, which reflects the light in many different ways. Tiny gold accents, in forms of things like birds or bees, called gold *pallions*, were sometimes applied, to be fired into the color below, and then a clear enamel was fired on top. The reason that women were the predominant enamellists at the turn of the century is that they had the dexterity to perform such delicate and intricate tasks. At the Robbins Company many women in the enamel department came from England, Sweden, and Italy, and enamel training was offered at the factory.

The 1908 "Cloisonne Sterling and Enamel Catalog" by the Charles M. Robbins Company is one the earlier full color jewelry catalogs by an Attleboro maker. Through information provided by the Robbins Company, it is known that the original Charles M. Robbins traveled to China in 1891 and spent six months studying the enameling process. Robbins was such an admirer of President McKinley that in 1892 he opened a workshop in his garage to manufacture enamel campaign badges for the McKinley election. The Robbins Company has a long history of enamel production and at one time operated the largest enameling department in the world in Attleboro.

A Simmons catalog, c. 1912, states "we already possess nearly fifteen thousand (15,000) different chain patterns and are continually adding more." Watch chains were among the primary items of jewelry for men and women in the early 1900s, keeping the owner's pocket watch secure and also making a fashion statement. Many Attleboro jewelers were chain manufacturers.

The Baer & Wilde Company introduced the "Kum-a-part," a separable cufflink, and enjoyed tremendous success. Their advertising was "a snap to button" and "here's the button for your soft cuffs" in magazines showing their cufflinks with color pictures of Fifth Avenue in New York, c. 1923. The company invested $750,000 in advertising between 1918 and 1930. Baer & Wilde offered "snappers," another term for these snap links, in enamel, celluloid, and pearl on gold or silver metals, as well as in sterling silver, solid gold, and platinum.

Although the separable cufflink was most popular in the 1920s, it had earlier beginnings. W.H. Wilmarth & Co. advertised "separable, lever and automatic sleeve buttons" in 1910. This book illustrates two examples of Victorian gold electroplate separable cufflinks, c. 1890. Other Attleboro firms offered their own versions of snap links in the 1920s, including J.F. Sturdy's "Stalokt," specially designed with a button, which is pressed to open the cufflink. The Bliss Bros. made Art Deco enamel snap links, considered scarce; one pair is included in this book.

Mesh dates to the 15th century with warriors wearing mesh skirts for protection. Fashion mesh purses from Attleboro had their beginning in the late 1800s, and at this time, the chain mesh was all hand made. The Whiting & Davis Company of Plainville created the first machine for the manufacture of mesh in 1909. Whiting & Davis and Mandalian are known for their exceptional enamel mesh purses from the 1920s and 1930s. The Mandalian Company of North Attleboro constructed a mesh screen for viewing movies at the local community movie theater. Movie stars traveled to Whiting & Davis in Plainville to be photographed with company president Charles A.Whiting and to receive mesh purses as gifts. Movie stars also endorsed Whiting & Davis products in magazine advertising, and mesh purses were among the utmost accessories to own. My grandmother Ethel Berry was given a Whiting & Davis white mesh purse by her husband William around 1920. My grandparents N. Mildred and B. Bertram Eldridge gave my mother, Patricia, two child's Mickey Mouse enamel purses by Whiting & Davis during the 1930s. It would seem that any woman would have been overjoyed to receive a Whiting & Davis or Mandalian enamel mesh purse as a gift, and these beautiful purses enjoyed tremendous popularity for many years.

Whiting and Davis Company and the Plainville Stock Company, Plainville, c. 1909.

Other Attleboro companies also produced fine mesh purses, although far more rarely seen than Whiting & Davis and Mandalian products. Paye & Baker made a small mesh purse in the shape of a fish. The Thomae Company made mesh purses in sterling enamel with compact tops. The Saart Bros., well known as makers of sterling enamel guilloche vanity items, also made purses. The Evans Case Company made mesh purses, sometimes with a compact and mirror attached, but these were manufactured for a only brief time, as the Whiting and Davis Company said they were too close in design to their own purse patents.

Lady's compacts and vanity cases in enamel, c. 1920-1940, are a beautiful part of Attleboro's history. The firm of George L. Brown made scenic guilloche enamel compacts, picturing such idyllic vistas as a country house with the sun rising and a river running alongside, where the sun and water actually appear illuminated by the translucent enamel process.

The J.M. Fisher Company is known for colorful, high style Art Deco enamel compacts. Ripley and Gowen-LaMode specialized in hand painted enamel compacts, often in sterling vermeil, and sometimes with a handle at the top, so the tiny compact had the appearance of a purse. The Finberg Co. compact in this book shows an enamel bluebird. Other Attleboro compact manufacturers include the D.F. Briggs Co., James E. Blake Co., George Webster, Thomae, Marathon, and Whiting and Davis.

The Evans Case Company operated in North Attleboro, Massachusetts from 1922–1960, and was a manufacturer of high quality enamel compacts, cases, lighters, vanity cases, and jewelry. Their products were advertised in fashionable magazines and sold worldwide.

Numerous firms still manufacture jewelry in Attleboro and North Attleboro, Massachusetts, today, some in business since the 1800s. This book shows the evolution of design from 1780-1960, as well as the quality and diversity of Attleboro jewelry and related items. Virtually every item in this book is stamped with the manufacturers' trademark or has been documented through old catalogs and advertising. The trademark section includes nearly 200 different jewelry manufacturers operating in the Attleboro area.

Many individual Attleboro jewelery manufacturers with histories from the1800s and early 1900s have created the jewelry in this book. Buttons have long been a form of personal adornment, and Eric W. Reichardt, an expert in this field has provided an excellent chapter on Attleboro buttons, to begin the history of jewelry from Attleboro, Massachusetts, known worldwide as the birthplace of American jewelry manufacturing.

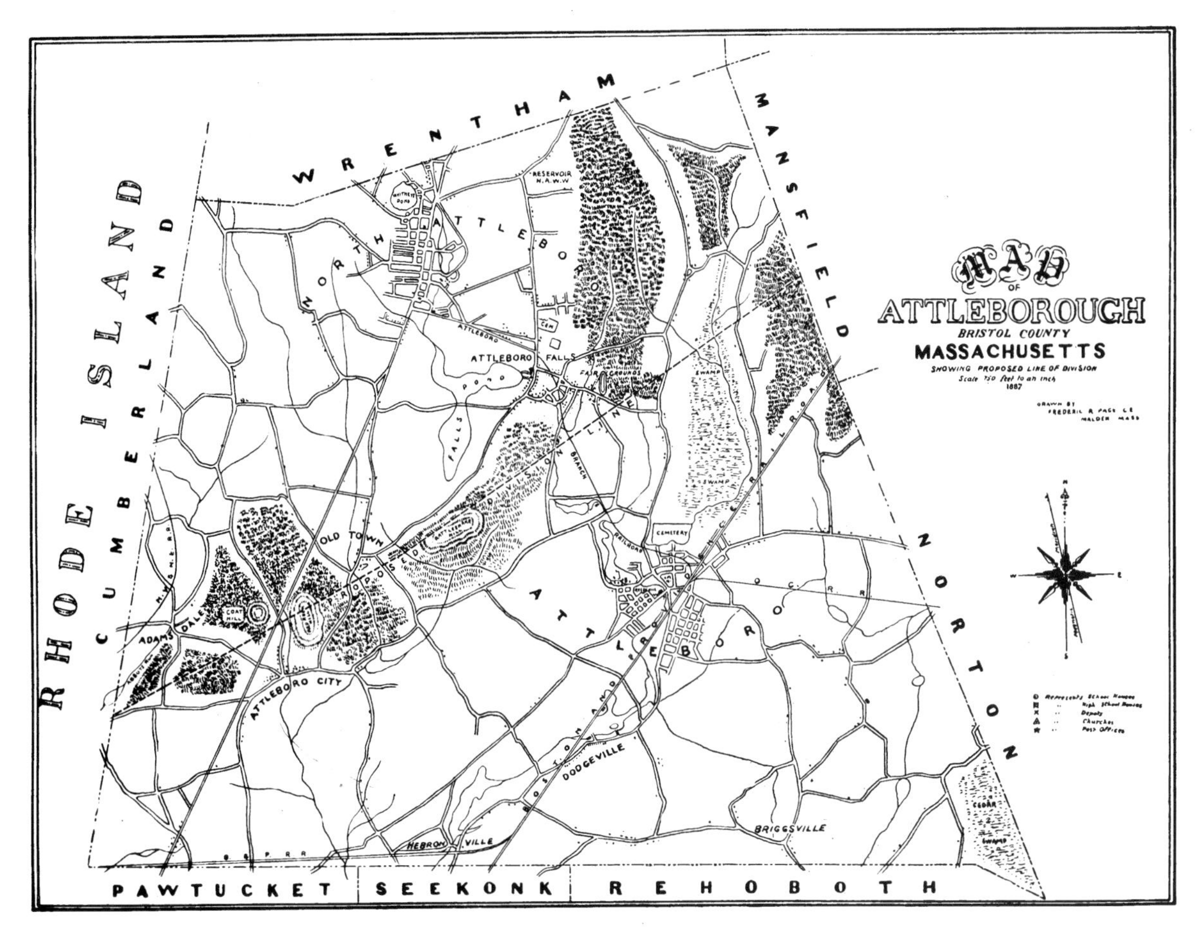

1887 map of the Attleboro, Massachusetts area.

Monument Square, Attleboro, c. 1915.

James E. Blake Co., Attleboro, c. 1908. Maker of sterling and *Sterline* jewelry, match safes, compacts, and dresser ware.

Watson and Newell Company, Attleboro, c. 1910. This company manufactured sterling enamel jewelry and souvenir spoons.

Residence of jewelry manufacturer C. Ray Randall, North Attleboro, C. 1915.

234 *REHOBOTH AND ATTLEBORO.*

town contained about thirty-two hundred people. There was a factory at Farmer's Village, and one at Falls Village. There was a mill at Dodgeville, and the town had three hundred and fifty looms and thirteen thousand spindles. At Robinsonville, about half way between the north and east villages, the button business had been started, and has since developed into considerable proportions. A shuttle shop with twelve men was in operation at East Attleboro; and — this is the first which we hear of Attleboro's jewelry — there was a very little jewelry business. In 1835 the Boston and Providence Railroad was built, and Attleboro village took a start and rapidly caught up with its sister village.

ONE OF THE FIRST JEWELRY SHOPS.

During the civil war the town did its full share in furnishing men and money and in home work. It contributed a company each to the Seventh, the Forty-seventh, and the Fortieth regiments, quite a number of men to the Twenty-fourth and Fifty-eighth, and had men in all branches of the army and the navy. Rehoboth was not behind at this time, her sons also being found in all departments of the service.

Attleboro grew rapidly after the war, in consequence of the jewelry business, till in 1886 it numbered fourteen thousand people. Unfortunately its growth was not concentrated; there were two large villages, four miles apart, nearly equal in all respects, each with a high school, a fire department, a water supply system, and everything pertaining to a large and prosperous town. The idea of a separation was discussed for some time before it took place, which was in July, 1887. At that date Attleboro, the east village, was made a town with the old name, with a population of six thousand nine hundred, a valuation a little over $3,000,000, and one thousand six hundred

One of the early jewelry shops in Attleboro from *New England Magazine,* 1907.

JEWELRY FACTORIES.

Jewelry Factories in Attleboro from *New England Magazine,* 1907.

The O.M. Draper factory in North Attleboro from *New England Magazine*, 1907.

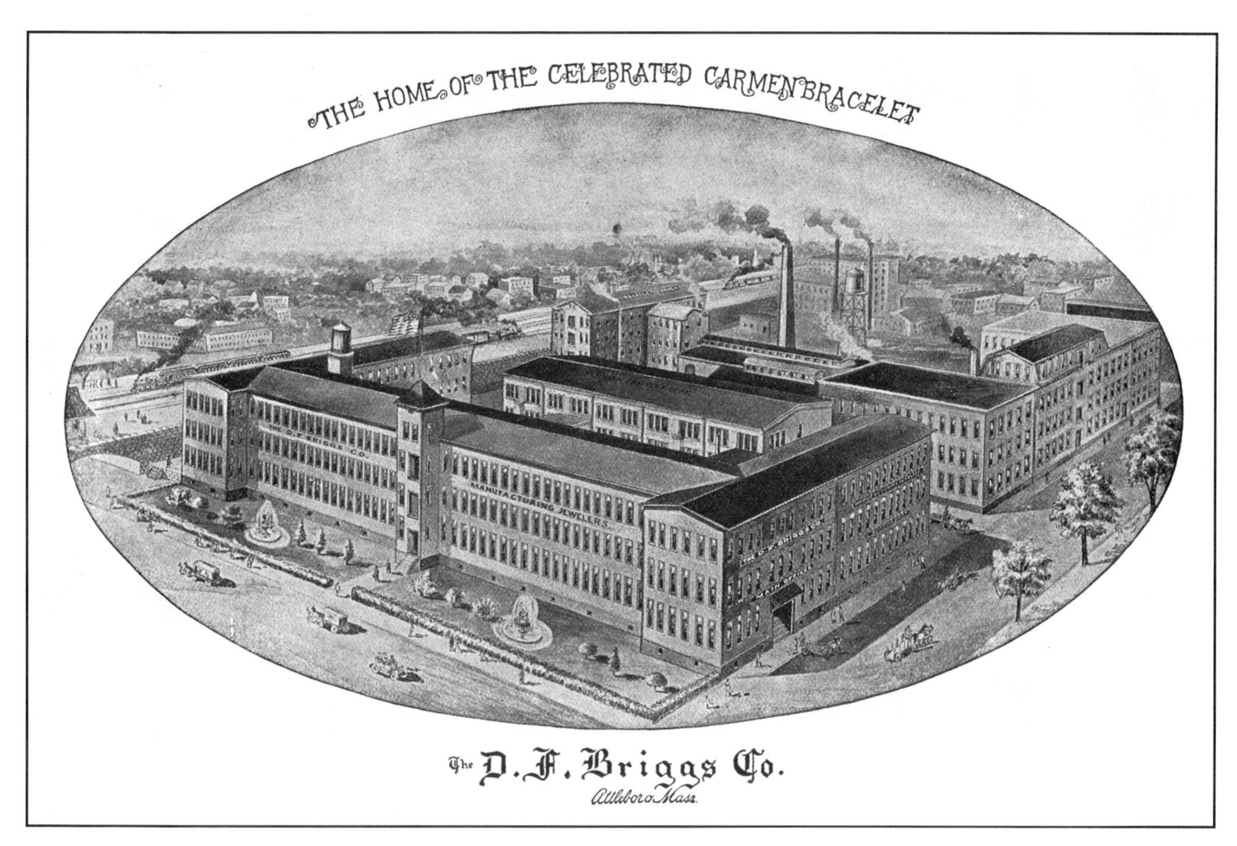

D.F. Briggs Co., Attleboro from *New England Magazine,* 1907.

Makepeaces's Factory in 1903 was also occupied by Allen, Smith & Thurston, C.H. Allen & Company, and George L. Brown & Company. Many firms shared manufacturing in Attleboro.

Regnell, Bigney & Company

Horton, Angell & Company

The James E. Blake Company

Chapter 2

History of Attleboro Button Production

by Eric W. Reichardt

Up until the Renaissance, loose-fitting cloaks were fastened by the use of a decorative pin or brooch, and other articles of clothing were fastened by use of either lace or hook and eye arrangements. The aristocracy, on the other hand, wore buttons of gold or silver inset with jewels, which closely resembled contemporary jewelry styles, though these were purely ornamental in nature and not functional "buttons" as we know of them today.

Certainly by the 16th century very small spherical buttons were in general use both to decorate and fasten together the doublets being worn by men at this time. Since then, decorative buttons have had an important impact in both fashion and society.

The button industry in the United States traces its known beginnings back to at least the 1720s when Caspar Wistar, a German immigrant, began making brass buttons in Philadelphia. An industrious young man, Wistar advertised his buttons for sale in local newspapers that came with a seven-year warranty against defects or breakage. Actual documented examples of these buttons were handed down through the generations in the Wistar family who donated a few to the Philadelphia Museum of Art, which are currently on public display.

During the 18th century, a majority of the buttons being worn by Americans were being imported from other countries like England and France. At the onset of the American Revolution, a need arose for uniforms and buttons for the newly-formed American Army. Even though we continued to import buttons from France, our imports from England dwindled, and new sources were needed to provide such goods. Thus many new industries arose in the colonies out of necessity, including the manufacture of clothing buttons.

Buttons being produced locally during this period were quite unsophisticated and made of cast pewter in either crude two-piece soapstone molds or small hand held scissor molds that could make only from 1 to 12 buttons in each pouring. Copper alloy products such as old ship sheathing, broken bells, brass pots and teakettles were also melted down to produce brass buttons since most of the early copper mines in the colonies like those located at Simsbury, Connecticut, or Belleville, New Jersey, had already become defunct by the onset of war with England. Up until this time Parliament had enacted strict laws prohibiting the colonists from the actual smelting or final production of copper alloy products, and all copper ores mined here had to be sent back to England for refining, although some rebellious colonists disregarded these imposed laws.

When the Revolutionary War ended, trade resumed with England, but we realized more than ever that we would have to become more self sufficient, and so the real beginnings of the button industry here in the United States began with various entrepreneurs starting up small family-run shops.

The earliest documented button producer from Attleboro, Massachusetts, was a man named Edward Price. An immigrant from Birmingham (England's center for button production during the 18th century), Price brought with him the necessary button tools and experience to begin to produce and sell finished buttons out of Attleboro in 1793. At this time, England was still the leader in fashion throughout Europe, and the "dandies" of aristocratic society were wearing very large fancy buttons the size of silver half-dollars (approx. 30-36mm). Most likely the type of buttons he produced would have been stamped out from imported rolled copper planchets and then decorated and finished, and would have been similar to the example shown in figure 1. No examples of buttons with an Edward Price maker mark are known, since at this time it still was not standard practice, though sometimes buttons are seen with a quality mark indicating how they were made, such as "plated" or "gilt." By the mid- to late 1790s, manufacturer's maker marks started to be stamped on button backs by some producers, particularly in England, though there are some known American makers who were located in the larger cities in the northeast who also marked their buttons by this time. By the early 19th century, not only quality and maker marks are seen, but also clothing outfitter and merchant names began to appear on button backs. These names help considerably in the dating of metal buttons produced after 1800.

Since the U.S. resumed importation of finished products from England after the Revolutionary War, it became very difficult for small enterprises in the U.S. to be competitive, and so Edward Price was forced to work for a newly established jewelry firm owned by David Brown and Obed and Otis Robinson, also out of Attleboro (this firm was established in 1807). Since Price's expertise in the button-making field was considerable, he eventually taught the Robinsons this field of endeavor.

The Robinson firm not only produced buttons under contract with the United States Army at the start of the War of 1812, but also made glass buttons for the civilian market. By 1820 this firm was being managed by Obed Robinson's son Richard and was called Richard Robinson

& Company. This firm became a major producer in this country of various metal buttons for both the military and civilian markets (see figures 2 & 4).

In the beginning, the Robinson firm ran their business in a small building with machinery powered by horses, but by 1827 the company had expanded and had a more substantial factory powered by water from the Ten Mile River. The company employed approximately seventy- five workers capable of producing up to one hundred gross buttons a day and their annual expenditure was up to $15,000 just for the natural gold ore used in the gilding process. The firm also had merchants and agents located in many of the larger cities, including Portsmouth, Boston, Providence, Philadelphia and Baltimore and also exported their finished products to foreign countries in South America , Haiti, and other countries in the West Indies.

In 1828 the Robinson firm acquired a new partner, William Henry Jones, who had practical experience from working in Birmingham, and so the firm took on the name of Robinson, Jones & Company (see figures 3, 6 & 10). By 1834 Jones left the firm to start his own factory in Waterbury, Connecticut and the Robinson firm was renamed R. & W. Robinson after Richard and Willard Robinson.

From 1834 to 1845 the R. & W. Robinson firm employed about one-hundred workers and made various metal buttons, including civilian, sportsmen, political, academic, uniform and military types. This firm was awarded high honors for their manufacturing expertise on numerous occasions by institutes in Boston, New York, and Philadelphia. Copies, or tokens, were made at the Robinson firm from these original silver medals, which were readily accepted by the general public as a medium of exchange since at this time coinage was scarce (see figure 17). The artistic quality and fine die work seen on early Robinson-made buttons has never been surpassed by any other American button manufacturer and the wide range of buttons produced make their products highly sought after by button collectors from all areas of this collecting field (see figures 7, 9 and 11-13).

During the mid 1840s, the Robinson firm, known for its fine fancy gilt buttons, became bankrupt due to changing clothing styles in the civilian market place and because of the Great Depression of 1833-1844, which caused the demise of many businesses throughout the country.

Fortunately, a former bookkeeper of the Robinson firm, Daniel Evans, saw this opportunity to continue this trade and bought out the factory holdings and surplus stock around 1848. During the 1850s, D. Evans not only sold off old surplus but continued to produce new designs using earlier Robinson-marked button backs, so for this reason, the earlier Robinson firm name does appear on certain buttons made up until the mid 1850's.

England's influence on the world of fashion had always been an important one, and after the death of Prince Albert in 1861, Queen Victoria began wearing jet buttons and jewelry, which helped promote the fad of darker colored buttons. Bright gold buttons gave way to buttons of black glass and dark enameled examples. Thereafter civilian buttons followed jewelry styles of the Art Nouveau and later Art Deco movement.

Since 1848, with the declining market for fancy gilt civilian buttons, the D. Evans firm had to rely more and more on the production of only gilt uniform and military buttons in order to survive. Before and during the Civil War, Evans was a major supplier of numerous uniform button types, some of which are represented in figure 16. Since then the D. Evans button manufactory has had a long and successful history in the marketplace and it continued throughout the rest of the 19th century to be one of the four largest producers in this country of various types of plated and gilt metal buttons.

In 1945, the Waterbury Button Co. of Waterbury, Connecticut, bought out the D. Evans button manufacturing facility, thus ending the long and prolific history of button production in Attleboro, Massachusetts (Also see figures 8, 14-15).

Button Chronology ***(working dates of the manufacturers)***

NOTE: The history and chronology of button makers mentioned in this article is based on metal buttons used in the conventional sense. It does not include the multitude of jewelry manufacturers of shoe and glove buttons; cuff-link buttons; or collar and cuff stud buttons which became prevalent during the late Victorian period.

Edward Price (1793-1811). Button maker from Birmingham, England, who moved to Attleboro in 1793 at the age of seventeen, bringing with him various tools of the trade from his previous experience over in England. The earliest documented button maker from Attleboro, Massachusetts, he later helped the Robinsons get started in the button manufacturing business in 1812. His son, George Price, later became a manufacturer of fire gilt jewelry with his initial partner, Calvin Richards, and later partner, S. S. Daggett. This was the third such establishment built in the Attleboro area.

Obed Robinson (c.1807-1820). Obed Robinson was both a blacksmith and colonel of the local militia in the Revolutionary War, who produced gunlocks for weapons used by the Continental Army. After the war, he produced clocks, and by 1807 he had opened a jewelry manufacturing shop with a very skilled workman named David Brown. In the War of 1812, he began to produce uniform buttons for the United States Army with the help of Edward Price and his three sons, Otis, Richard & Willard. He later retired before his death in 1840.

Richard Robinson & Co. (c.1820-1834). By 1820 Obed Robinson's son, Richard, became the general manager of the button firm. Military contracts for "bullet" or "ball" buttons were made along with Artillery Corps (AY55[1]) and militia uniform buttons. Also "dead-eye" glass buttons were produced during the 1820s. The "R.R. & Co." b/m[2] is also seen on patriotic, political and fancy civilian buttons.

Robinson, Jones & Draper (1826-c.1830s). Button firm of Richard & Willard Robinson, William H. Jones and H.M. Draper. Jones left the partnership in 1834 (see also Robinson, Jones & Co.).

Robinson, Jones & Co. (1828-1834). Partnership until 1834 when W. H. Jones left the firm to start his own button manufacturing facility in Waterbury, Connecticut. This b/m is seen on both civilian and uniform buttons.

Robinson, Blackinton & Co. (1831-1835). Virgil Blackinton, born May 12, 1796, married Hannah, daughter of Obed Robinson and was involved in the manufacture of buttons with Richard and Willard Robinson. This b/m is seen on both civilian and one Navy uniform button (NA86)[3].

Robinson, Hall & Co. (1832-1836). A shop was set up in an old cotton mill on the Seven Mile River near Newell's Tavern in West Attleboro. They made plain metal buttons for the civilian market and by 1837 the manufacture of jewelry was commenced by W. H. Robinson (This firm also had a shop located in Old Town with partner S. L. Daggett which manufactured jewelry). In 1843, J. H. Hodges and J. T. Bacon bought this facility to continue the manufacture of jewelry, and it is said that they were the first in town to actually electroplate brass jewelry. They continued in business until 1847.

Edward Hulseman (1833-1836). Edward Hulseman was a diesinker or engraver for the Robinson button manufacturing firm. His maker mark "H" is seen under the word "New York" on both the 1833 and 1836 tokens shown in fig. 17. "Hard Times" tokens were issued during the Depression in the U.S. from 1833 to 1844. These copper tokens, the size of a U.S. large cent, freely circulated as a medium of exchange. The Robinson firm is credited with manufacturing not only the two aforementioned tokens but numerous others (see also HT25, HT70, HT152-156 and HT428).[4]

R. & W. Robinson (1835-c.1845) This button firm, made up of the two sons of Obed Robinson (Richard and Willard), continued under this name even after Richard died in 1838. They made numerous varieties of both civilian and uniform buttons. This company was nationally recognized on at least five occasions for its outstanding workmanship seen on numerous button designs. The factory holdings were bought out after their bankruptcy by Daniel Evans around 1848.

Willard Robinson & John Hatch (1845-1849) John Hatch, a mechanic working for Willard Robinson, designed and patented an automated button machine (Patent No. 3,915., Feb. 20, 1845).[5] The brass, and later tin suspender buttons were made at the rate of 23 per minute. Six of these machines were made and subsequently one of them was sold and exported to Germany. On Nov. 3, 1863, Robinson reapplied for an extension on the original patent and obtained a government contract to produce these "Hatch pantaloon" buttons for Union troops during the Civil War. The earliest Hatch buttons produced during the 1840s had a Robinson b/m and were made of brass. Willard Robinson's son Arthur continued this business until about 1893, using one of these original machines though suspender buttons by this time were quite different and mostly made utilizing newer machinery and techniques.

Draper & Sandland (1845-1853) This firm produced both Navy and Boston City Guard uniform buttons (NA101; NA107; MS57).[6] A brass calendar token from this firm dated 1853 advertised "Manufacturers of Every Description of Gilt Buttons / Manufacturers & Dealers in Gilt & Plated Jewelry". Draper continued after this partnership dissolved to produce tortoiseshell jewelry in Mansfield, Massachusetts during the 1850s.

Freeman & Co. (1846-1878). Benjamin Freeman founded the business in 1846 with about twenty-five workers and made rings and other plated wares. The company became Freeman Brothers (Freeman & Bro.) in 1849, when Benjamin's brother Joseph became a partner. The firm was renamed Freeman Brothers & Co. in 1855 when Virgil Richards became a partner. In 1860 the firm became Freeman & Co., and during the Civil War the company made Marine uniform buttons (MC10)[7] along with corps badges and brass neck chains for the soldiers. It became B. S. Freeman & Co. in 1879, when Joseph J. Freeman died, and they made mostly chains and bracelets thereafter.

D. Evans & Co. (c.1848-1945). Daniel Evans worked for, and finally took over, the R. & W. Robinson firm after their bankruptcy in c.1845. The firm made both civilian and uniform buttons. It was considered one of the four largest button manufacturing facilities in the United States and had the longest running production of all the known makers out of Attleboro. It went out of business when the Waterbury Button Co. of Waterbury, Connecticut bought them out in 1945. Numerous civilian and uniform buttons are seen with the D. Evans b/m, the most desirable being of pre-1866 vintage.

The following backmarks of clothing outfitters and merchants were also made by the D. Evans firm: [8]

Thomas Cahill, Boston (c.1850s)
Thomas N. Dale, N.Y. (c.1861-1882)
John Earl Jr., N.Y. (c.1850s)
E.L.E. & Co., London (c.1860s-1870s)
Frederick F. Hassam, Boston (c.1861-1870)
Macullar, Williams & Parker, Boston (c.1851-1879)
Schuyler, Hartley & Graham, NYC (c.1854-1878)

[1]Alphanumeric codes used in chronology section are from the following source: *Record of American Uniform and Historical Buttons*, by Alphaeus Albert (see p.56 for photo of this example).
[2]b/m is the abbreviation for back mark. This manufacturer's stamp is seen on the back side of some buttons.
[3]See p.99 in Albert's book for photo
[4]See p. 95 in *Standard Catalog of United States Tokens 1700-1900*, by Russell Rulau.
[5]The original patent including complete description and sketches may be seen at the U.S. patent office.
[6]See p.101 & 171 in Albert's book for photos
[7]See p.110 in Albert's book for photo
[8]See p.25 in *Uniform Buttons of the United States*, by Tice (Some dates from this source are modified).

Figure 1: One-piece, c. 1780s-1790s silver plated coat button; b/m: "PLATED"; 31 mm. $50-150.

Patriotic Buttons

Figure 2: Patriotic or Military, c. 1813-1820s (UU132); b/m: "R. R. & Co."; 15 mm. $50-85.

Figure 3: Patriotic, motto "In God We Trust" c. 1828-1834; b/m: "Extra Rich / R. J. & Co."; 15 mm. $30-50.

Figure 4: Congressional Campaign of 1834, motto "Whigs of 76" (PC97); b/m: "R. R. & Co."; 12 mm. $125-175.

Fancy Civilian Buttons

Figure 5: One-piece fancy gilt, c. 1820s-1830s; b/m: "Robinsons / Treble Gilt"; 21 mm. $10-30.

Figure 5: One-piece fancy gilt, c. 1820s-1830s; b/m: "Robinsons / Extra"; 16mm. $10-30.

Figure 5: One-piece fancy gilt, c. 1820s-1830s; b/m: "Robinsons / Extra"; 17mm. $10-30.

Figure 6: One-piece plain gilt, c. 1828-1834; b/m: "Robinsons, Jones & Co // Treble Gilt"; 18 mm. $3-8.

Figure 6: One-piece fancy gilt, c. 1828-1834; b/m: "R .J. & Co / Extra Rich"; 17 mm. $10-30.

Figure 6: One-piece fancy gilt, c. 1828-1834; b/m: "R. J. & Co / Extra"; 17 mm. $10-30.

Figure 7: Two-piece fancy gilt, c. 1835-1848; b/m: "R & W Robinson / Extra Rich"; 24 mm. $10-25.

Figure 7: Two-piece fancy gilt, c. 1835-1848; b/m: "R & W Robinson"; 21 mm. $10-25.

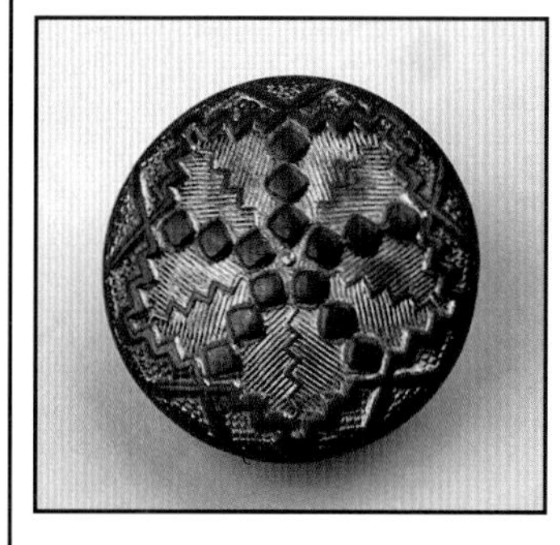

Figure 7: Two-piece fancy gilt, c. 1835-1848; b/m: "R & W R. / Extra Rich"; 20 mm. $10-25.

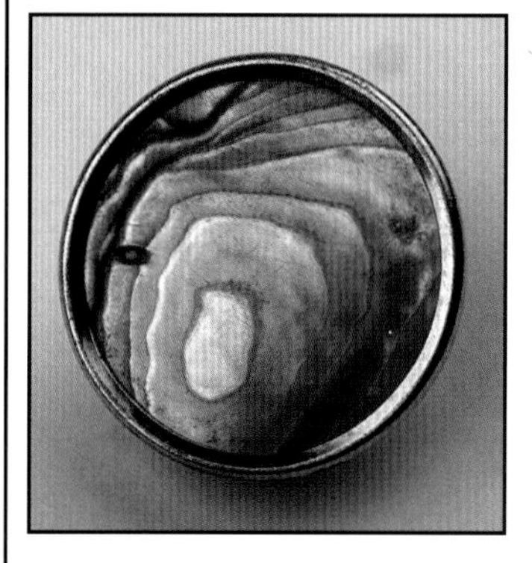

Figure 8: Mother of Pearl, c. 1850s; b/m: "D. Evans & Co. Attleboro Mass." (b/m in ribbon); 23 mm. $25-35.

Figure 8: Mother of Pearl, c. 1850s; b/m: "D. Evans & Co. Attleboro" (b/m in ribbon); 21 mm. $25-35.

Militia Uniform Buttons

Figure 9: One-piece Massachusetts Militia, c. 1835-1840 (MS26); b/m: "R & W Robinson / Attleborough // Super // Fine"; 22 mm. $60-85.

Figure 11: One-piece Artillery Corps, c. 1835-1840 (AY57); b/m: "R & W. Robinson / Attleborough // Extra // Rich"; 22 mm. $50-85.

Figure 10: One-piece Sarsfield Guards, Boston, c. 1828-1834 (MS95); b/m: "Robinsons / Jones & Co."; 22 mm. $300-375.

Figure 12: One-piece Vermont Militia, Light Infantry, c. 1835-1840 (VT7); b/m: "R & W Robinson / Attleborough // Extra // Rich"; 22 mm. $150-225.

Police Uniform Buttons

Figure 13: Police, generic, c. 1840s; b/m: "R & W Robinson / Extra Rich"; 23 mm. $30-45.

Figure 14: Police, generic, c. 1850s; b/m: "D. Evans & Co. Attleboro Mass." (b/m in ribbon); 23 mm. $20-30.

Figure 15: Police, Taunton, Mass., c. 1860s; b/m: "D. Evans & Co. / Attleboro Mass."; 23 mm, $10-20.

Civil War Uniform Buttons

Figure 16: Three-piece Massachusetts Militia, c. 1840s (MS35); b/m: "R & W Robinson / Makers // Attleborough// Mass"; 22mm, $30-50.

Figure 16: Two-piece Naval Officer, c. 1850s (NA112); b/m: "D. Evans & Co. / * Attleboro Mass. *"; 23 mm. $25-35.

Figure 16: Two-piece Virginia Military Institute, c. 1850s (SU408); b/m: "D. Evans & Co. Attleboro" (b/m in ribbon); 21 mm, $125-175.

Figure 16: Three-piece Staff Officer, c. 1860s (GS7); b/m: "D. Evans & Co. / Attleboro Mass"; 23 mm. $30-45.

Figure 16: Two-piece Infantry Officer, c.1861-65 (GI88); b/m: "D. Evans & Co. / *Attleboro Mass*"; 23 mm. $25-35.

Hard Times Tokens

Figure 17: Token: Robinson, Jones & Co –1833 (three known varieties); common circulated varieties worth $20-50; uncirculated examples command a premium ($225); rarest variety is silvered ($300 +); common variety shown in EF condition: $45.

Figure 17: Token: R & W Robinson – 1836 (three known varieties); common circulated varieties without dash between "New York" worth $20-30 and with dash between "New-York" from $45-200; uncirculated examples command a premium ($400); rarest varieties are of silvered copper ($450 +); example shown with dash is in VG condition: $50.

Chapter 3

Jewelry

Amethyst and gold bracelet by J.M. Fisher, c. 1920 marked "J.M.F. 14K", $500-750.

Edwardian bracelet by Daggett and Clap with glass stones to resemble old mine-cut diamonds and rubies, gold filled with filigree, c. 1908, $225-275.

Art Nouveau lady's lockets, gold filled with glass stones in the style of old mine-cut diamonds, c. 1905, W. & S. Blackinton and S.O. Bigney, $250-350 each.

Filigree pin by Finberg Manufacturing Co., gold filled with glass stone resembling peridot, c.1920 $100-150.

Art Nouveau chatelaine pin, c. 1910, by Simmons, gold filled, $75-125.

Victorian locket by W. & S. Blackinton, c. 1885, cockatiel on a fence, gold filled $175-225.

Victorian mechanical pencils by Aiken, Lambert, & Co. and H.F. Barrows, c. 1900, gold filled with engraving $75-100.

Victorian brooch by Plainville Stock Company, showing the Etruscan Revival influence with stamped patterns resembling gold granulation, and set with a cushion cut white stone, c. 1885. $200-250.

Black onyx cameo scarf pin or stick pin by the Plainville Stock Company, marked P.S. Co., c.1890. $175-225.

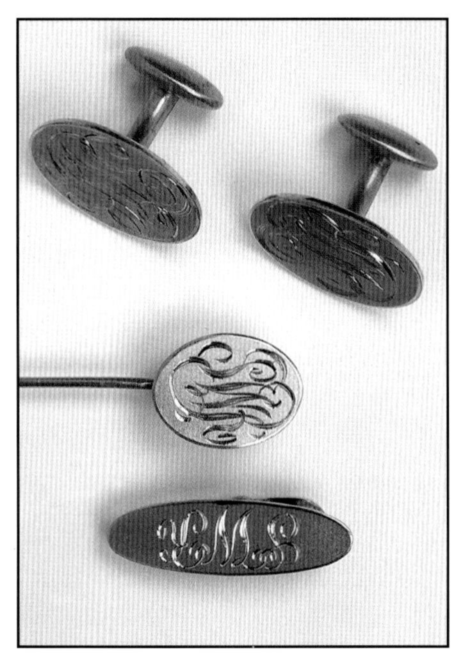

Stickpin, cufflink, and tie clip set by George L. Paine, gold-filled and hand engraved, c. 1900, $125-150.

Cuff links by Smith & Crosby, hand engraved 10K gold, patented 1880. $175-225.

Victorian bangle bracelets, gold filled with engraving, c. 1900 by J.F. Sturdy & Sons, marked "J.F.S.S." and George L Paine, marked "G.L.P." The latter example is marked "1/10" indicating a high quality of gold-filled jewelry. $175-225 each.

Victorian lockets, both by W. & S. Blackinton, gold filled with glass stones, c. 1900, $125-175 each.

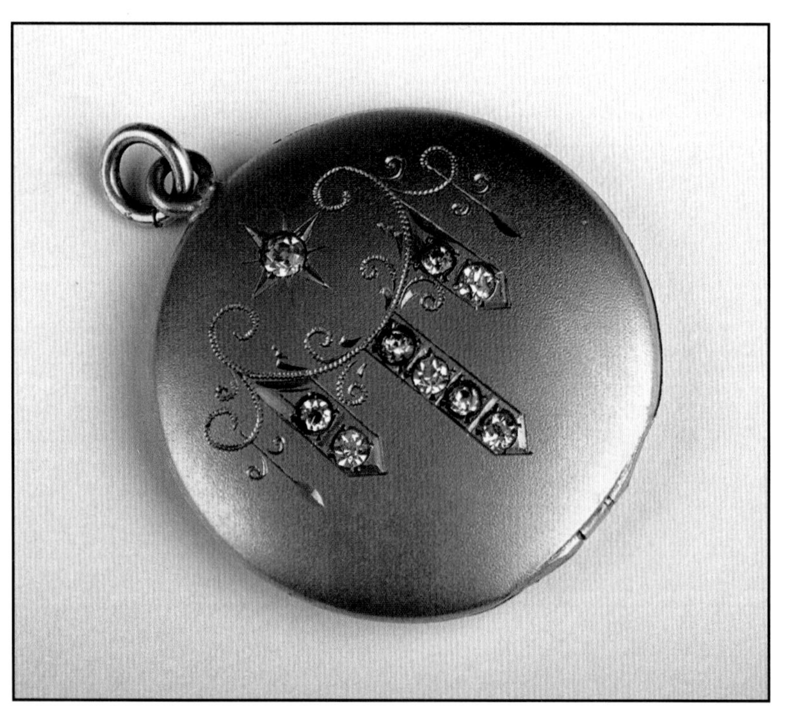

Locket by J. M. Fisher & Company, gold-filled with glass stones c. 1910. $150-225.

Victorian locket by J. F. Sturdy's Sons, c. 1895, marked "J. F. S. Sons, 14K Gold Stiffened". $125-175.

Victorian locket by R. B. McDonald, marked "R.B.M. 1874 Atrice" with a bee and patented "1903"; gold filled with glass stones, this large locket measures 2". $200-250.

Bangle bracelets by Hayward, engraved patterns with enamel, gold filled, $100-125.

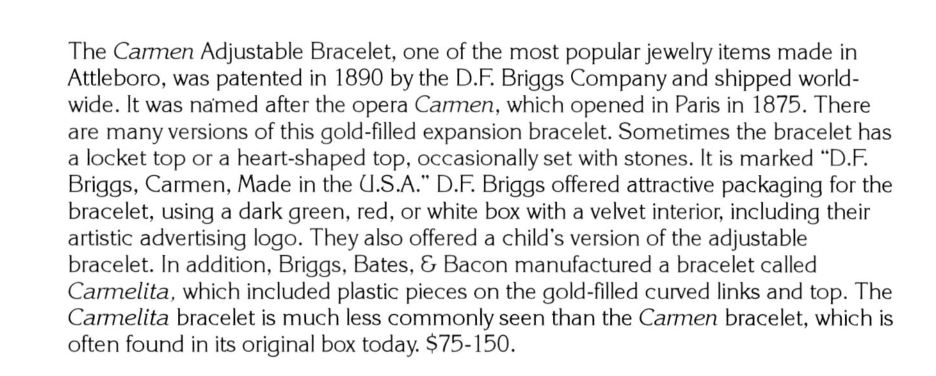

The *Carmen* Adjustable Bracelet, one of the most popular jewelry items made in Attleboro, was patented in 1890 by the D.F. Briggs Company and shipped worldwide. It was named after the opera *Carmen*, which opened in Paris in 1875. There are many versions of this gold-filled expansion bracelet. Sometimes the bracelet has a locket top or a heart-shaped top, occasionally set with stones. It is marked "D.F. Briggs, Carmen, Made in the U.S.A." D.F. Briggs offered attractive packaging for the bracelet, using a dark green, red, or white box with a velvet interior, including their artistic advertising logo. They also offered a child's version of the adjustable bracelet. In addition, Briggs, Bates, & Bacon manufactured a bracelet called *Carmelita,* which included plastic pieces on the gold-filled curved links and top. The *Carmelita* bracelet is much less commonly seen than the *Carmen* bracelet, which is often found in its original box today. $75-150.

The Carmen bracelet box, showing their logos on the exterior and interior.

American Queen adjustable bracelet by Pitman & Keeler. Other Attleboro makers of expansion bracelets included Mason, Howard, & Co., F.H. Sadler with their *Norma* bracelet, and the A.H. Bliss Company. $140-175.

The *Carmen Adjustable Bracelet* was advertised c. 1910 on this tin button, with a label on the reverse "The D.F. Briggs Co., Mfg. Jewelers, Attleboro, Mass. $75-100.

The *Midget Adjustable Baby Bracelet* was the name for a child's expansion bracelet offered by Fontneau & Cook Co., as depicted on this tin advertising mirror, c. 1905. The fleur-de-lis symbol shown in the center was a trademark of this company. $75-100.

Brownie stick pins, attributed to Regnell & Bigney. The charming Brownie figures were the design of Palmer Cox. Shown here are the Chinaman, Scotsman, and Police Cop, in enamel on brass. Brownie jewelry was illustrated in a Regnell & Bigney advertisement and shown in a booklet commemorating the 200th anniversary of Attleboro in 1894. $60-80 each.

A pen & ink drawing by Palmer Cox showing the Brownies.

Black cat stickpin, attributed to McRae & Keeler. This pin was advertised in their 1897 catalog. It is black enamel on brass with green glass eyes. The stickpin belonged to the author's grandmother. $60-80.

BLACK CAT JEWELRY.

"A Black Cat Brings Good Luck."—This saying has proved true in many instances, and at the present time the whole world are *quite* insane over this popular mascot. *A Red-headed Girl* and *The White Horse* are not in it. **"A Black Cat for Luck,"** is the old saying, and *it goes*. This craze cannot be considered a mere *fad* recognized by a certain class or classes. *This is a craze* recognized by everyone who seeks a *full portion* of this World's Goods, which may be obtained with more certainty by wearing **The Black Cat Mascot.**

We make them in STICK, SCARF, HAT and BROOCH PINS and LAPEL BUTTONS. These pins are true to life, the eyes are little emeralds, and their rich green brilliancy showing through the jet black head gives to the face a perfect expression.

These goods are all gold plated except the head, which is *jet black*.

LADIES' PRESENT.

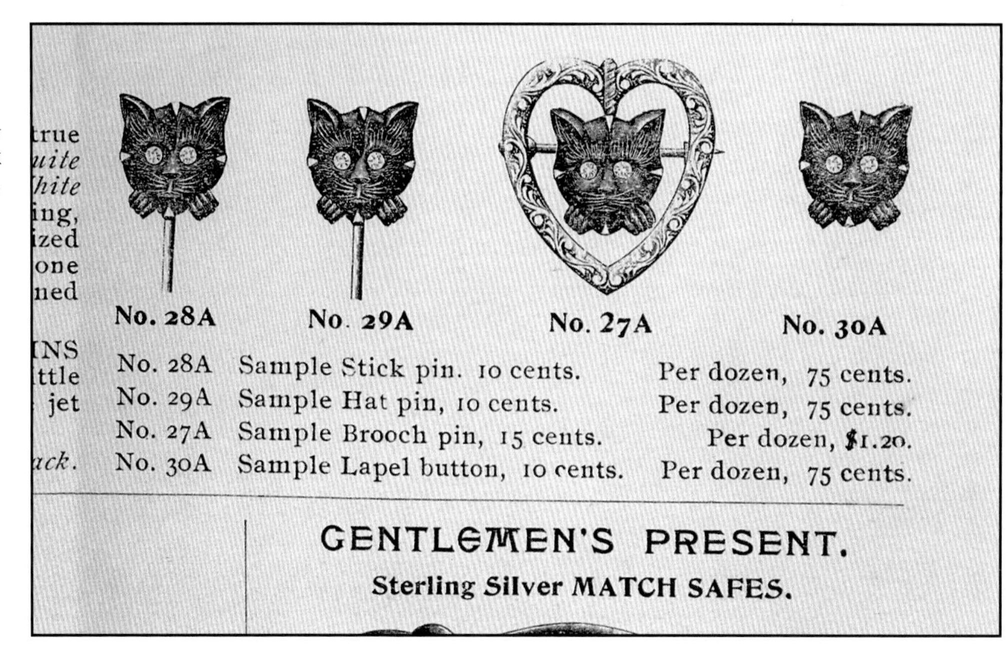

Advertisement from the McRae & Keeler catalog of 1897 for the black cat stickpin.

An advertisement in *The Jewelers Weekly*, December 29, 1886 for R.B. McDonald, in business in Attleboro c. 1874-1922 and famous for the manufacture of lockets.

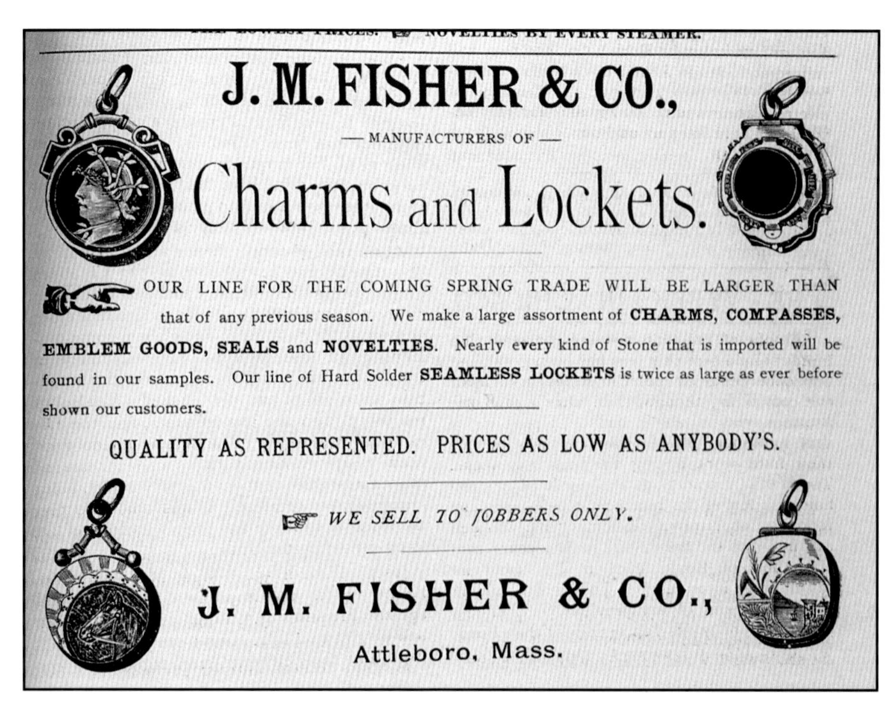

The J.M. Fisher Co. advertisement in *The Jewelers Weekly* of 1886. Other Attleboro makers who placed ads in the same issue were R.F. Sturdy & Co., C.R. Harris, Daggett and Clap, Bliss Bros. & Everett, W & S. Blackinton, and Plainville Stock Company.

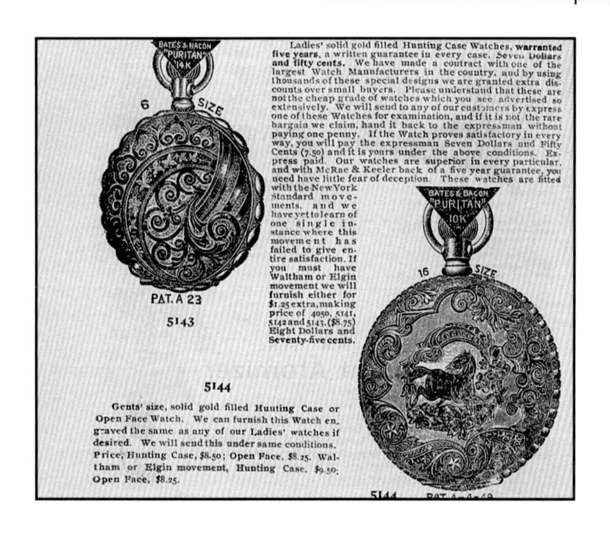

Pocket watches with cases by Attleboro maker Bates and Bacon are shown in the McRae & Keeler catalog of 1897, presented as gold-filled watches for ladies and gentlemen with Waltham, Elgin, or New York Standard movements.

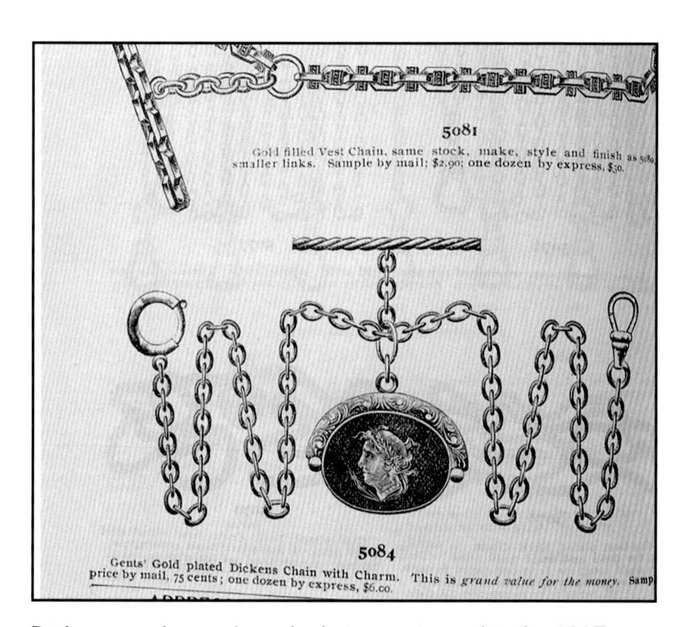

Dickens gentlemans' watch chain, as pictured in the 1897 McRae & Keeler catalog, is described as "*a grand value for the money*". The sample price was seventy-five cents, or six dollars per dozen by express. The charm appears to be a carved intaglio.

Simmons watch chain advertisement, c. 1905 offers "*12,000 patterns of men's vest, women's lorgnette and neck chains"*, showing the importance of specialty chains during the Edwardian era.

Chatelaine pins, for ladies pocket watches, from an early 1900s Simmons catalog.

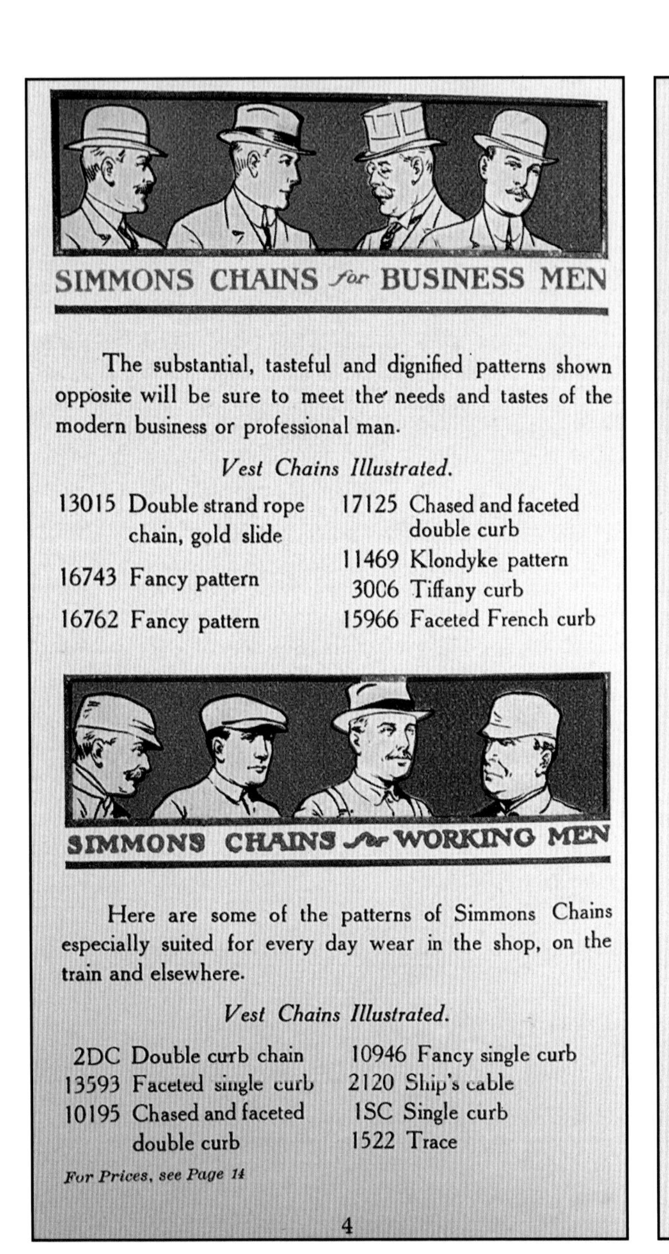

SIMMONS CHAINS *for* BUSINESS MEN

The substantial, tasteful and dignified patterns shown opposite will be sure to meet the needs and tastes of the modern business or professional man.

Vest Chains Illustrated.

13015 Double strand rope chain, gold slide
16743 Fancy pattern
16762 Fancy pattern
17125 Chased and faceted double curb
11469 Klondyke pattern
3006 Tiffany curb
15966 Faceted French curb

SIMMONS CHAINS *for* WORKING MEN

Here are some of the patterns of Simmons Chains especially suited for every day wear in the shop, on the train and elsewhere.

Vest Chains Illustrated.

2DC Double curb chain
13593 Faceted single curb
10195 Chased and faceted double curb
10946 Fancy single curb
2120 Ship's cable
1SC Single curb
1522 Trace

For Prices, see Page 14

4

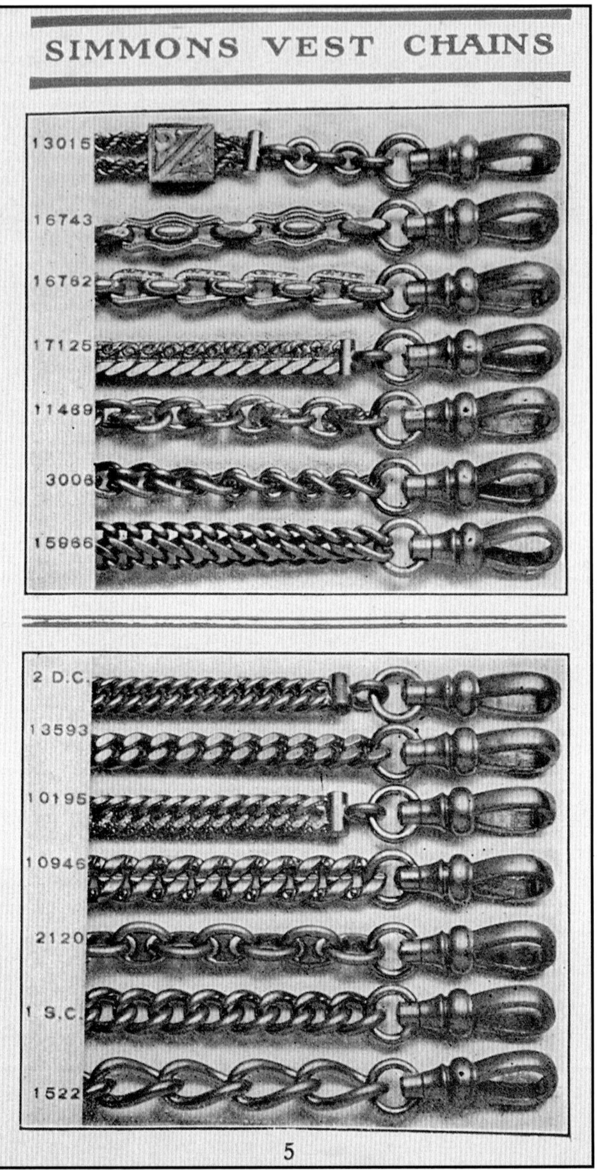

Simmons vest chains are pictured c. 1903 in a booklet titled *One Generation*, indicating the firm had been in business for about thirty years. Chains were offered for both businessmen and working men.

For all Women

The varied character of the patterns of Simmons Lorgnette Chains insure pleasing all feminine tastes—whether they incline to smartness or to quiet elegance in dress and adornment.

Lorgnette Chains Illustrated

12707/4232 Polished curb chain, polished slide, set with pearls.

9316/4250 Polished knurled cable chain, Roman front slide, set with ruby and pearls.

12439/3644 Roman cable chain, Roman slide, set with pearls and turquoise.

12079/4277 Polished cable chain, polished engraved slide, set with ruby and pearls.

12397/4198 Polished cable chain, polished engraved slide, set with pearls.

11721/4222 Polished cable chain, polished enamel slide, set with pearls.

12400/3609 Polished knurled cable chain, polished engraved slide, set with pearls.

11034/4274 Polished curb chain, Roman front slide, set with diamond.

5354/4181 Polished rope chain, polished embossed slide, set with ruby.

All lorgnette chains shown are 48 inches long and are gold soldered. All slides illustrated are 10K solid gold.

For Prices, see page 14

SIMMONS LORGNETTE CHAINS

12707 - 4232

9316 - 4250

12439 - 3644

12079 - 4277

12397 - 4198

11721 - 4222

12400 - 3609

11034 - 4274

5354 - 4181

7

Simmons lorgnette chains are advertised c. 1903 as 48" long, gold soldered, and with 10K solid gold slides. The style during Edwardian times was to wear the long chain with an attached lorgnette or watch, then to place the object in the pocket of a dress, for a draped effect.

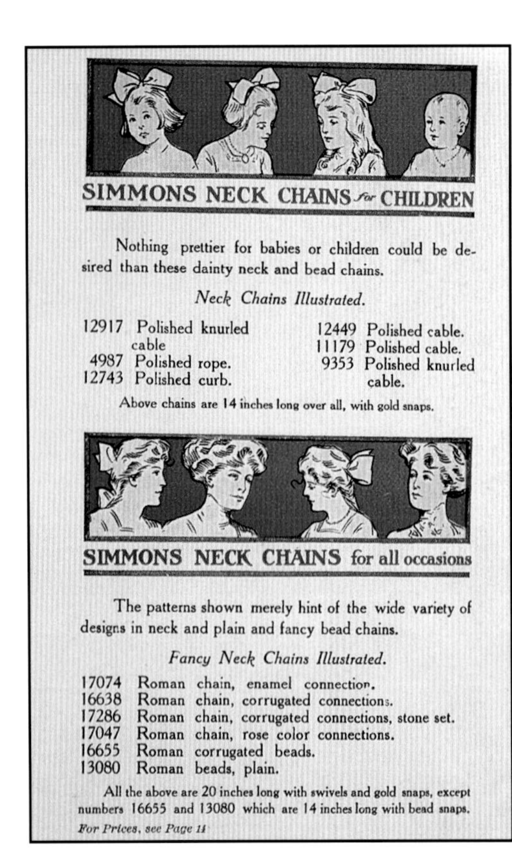

SIMMONS NECK CHAINS for CHILDREN

Nothing prettier for babies or children could be desired than these dainty neck and bead chains.

Neck Chains Illustrated.

12917 Polished knurled cable
4987 Polished rope.
12743 Polished curb.
12449 Polished cable.
11179 Polished cable.
9353 Polished knurled cable.

Above chains are 14 inches long over all, with gold snaps.

SIMMONS NECK CHAINS for all occasions

The patterns shown merely hint of the wide variety of designs in neck and plain and fancy bead chains.

Fancy Neck Chains Illustrated.

17074 Roman chain, enamel connection.
16638 Roman chain, corrugated connections.
17286 Roman chain, corrugated connections, stone set.
17047 Roman chain, rose color connections.
16655 Roman corrugated beads.
13080 Roman beads, plain.

All the above are 20 inches long with swivels and gold snaps, except numbers 16655 and 13080 which are 14 inches long with bead snaps.

For Prices, see Page 11

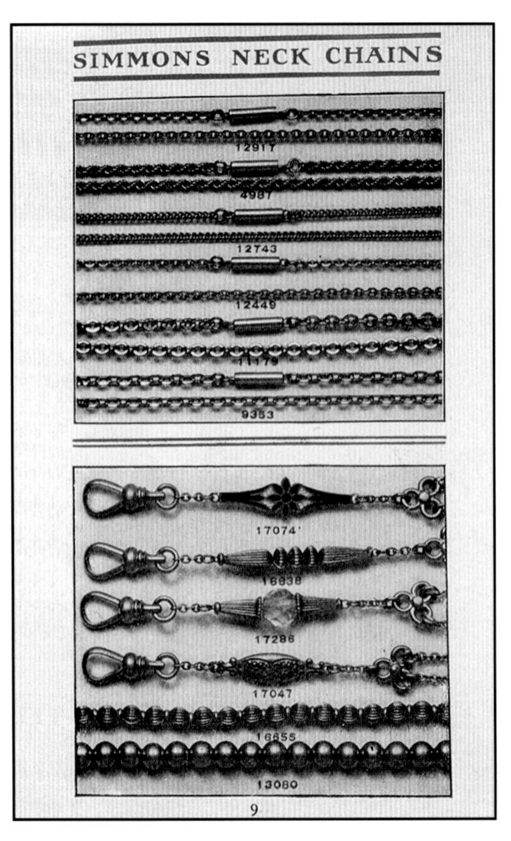

SIMMONS NECK CHAINS

Neck chains were offered by Simmons c. 1903 as "nothing prettier for babies and children could be desired than these dainty neck and bead chains." Women's chains are pictured with enamel and stones. Gold bead necklaces, both plain and patterned, were at the height of their popularity during the early 1900s.

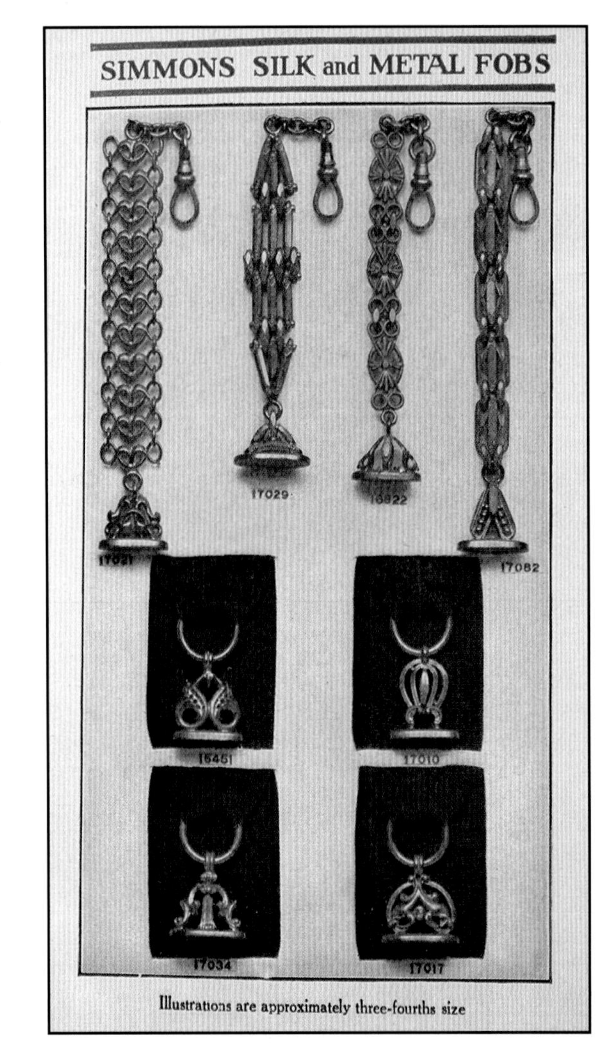

SIMMONS SILK and METAL FOBS

Illustrations are approximately three-fourths size

Silk and metal fobs were another way to carry a pocket watch, c. 1903. Simmons, and other Attleboro makers offered various styles, often personalized with the owner's initials.

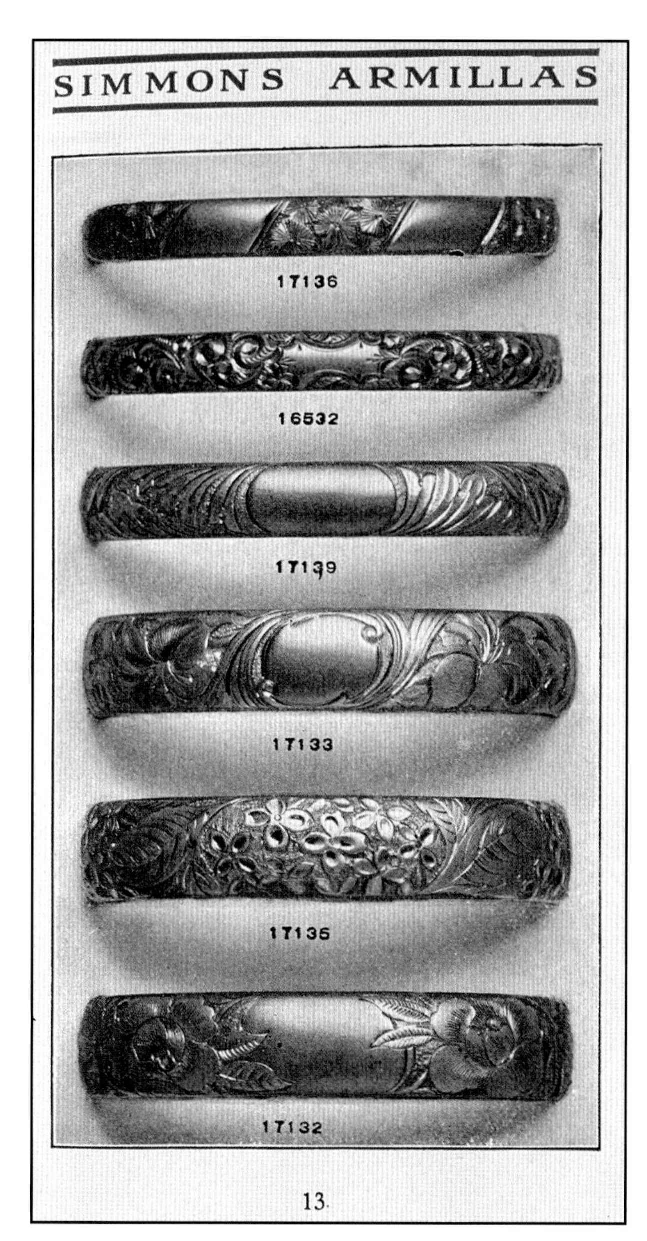

Many Attleboro firms manufactured bangle bracelets c. 1903. The Simmons *Armilla* was shown in a variety of designs, from plain to beautifully engraved, chased, and stone set.

The Robbins Company catalog 80E was published in 1919 and states "the articles pictured in this catalog are all manufactured in the United States of America by the Robbins Company, Attleboro, Mass., U.S.A. Makers of souvenir jewelry of all kinds, emblems, fobs, medallions, sport medals, and special order work of every description".

Watch fobs pictured in the 1919 Robbins catalog.

REO watch fob by the Robbins Company, opaque and iridescent translucent enamel on brass with original leather strap, c. 1925, "REO Motor Car Company, Lansing, Michigan, U.S.A." $150-225.

Sporting medals include a lady holding a laurel wreath in the 1919 Robbins catalog.

Sporting medal in sterling silver, as pictured in the Robbins catalog. Marked "Robbins Co., Attleboro", $75-100.

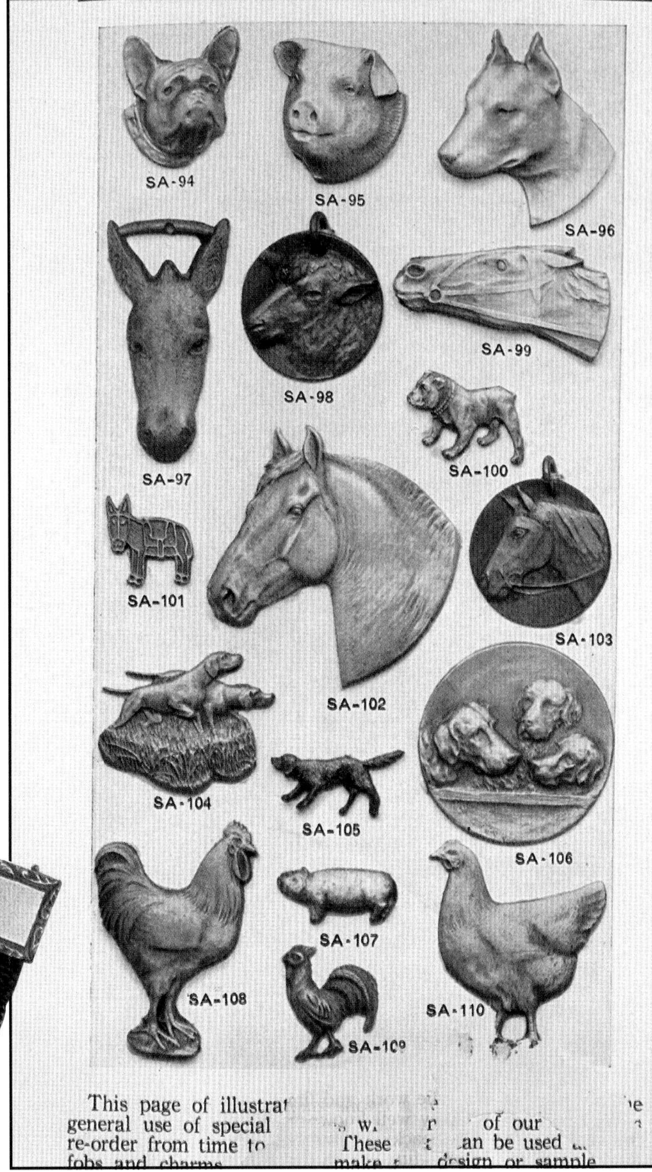

Animal designs were available as fobs and charms from the Robbins Company in their 1919 catalog.

Boxing Medal by Balfour, c. 1925, Chicago Tribune Golden Gloves Tournament. The Balfour Company is also known for college fraternity and sorority jewelry, trophies, and class rings. In 1919, Attleboro high school students wore class pins; in 1920 they wore class rings, both by Balfour; medal $40-50.

The Robbins Co. made badges for many businesses and associations, such as this one for an architects' convention in Atlantic City in 1926. $20-25.

Cross country track medal by Robbins Co., c. 1931. $40-50.

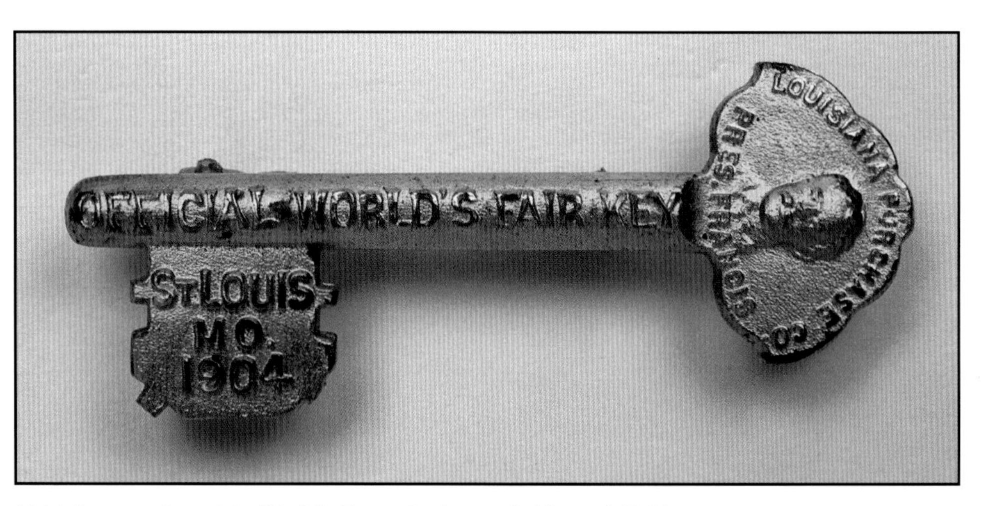

1904 Souvenir key of the World's Fair in St. Louis, Robbins. $40-50.

Craftsmen in Metal was a trademark of the Robbins Co., who issued this complimentary medal to advertise badges, medals, and souvenirs. $20-30.

Massachusetts Bay Tercentenary medal by Robbins Co. shows Puritan Governor Winthrop and Indian Chief Chickatabot in 1930. $15-20.

Century of Progress medal, c. 1933 by Robbins Co., who also made silver rings for this world's fair and many souvenirs for the 1939 New York Worlds Fair. $20-25.

Art Nouveau brooch by James E. Blake, c. 1910, marked on reverse "Sterling Front" and with a star punch mark, a Blake trademark. $175-275.

Art Nouveau graduation girl pin, c. 1910, with sterling front is typical of the work of James E. Blake. $50-90.

Art Nouveau hand mirror by James E. Blake in *Sterline*, an alloy resembling silver, c. 1910. $225-300.

Art Nouveau match box holder by the Mauser Manufacturing Company shows images of two women with cigarettes, early 1900s, sterling silver. $400-500.

Medallion pin by Frank M. Whiting, sterling silver. $250-325.

Bracelet by C. Ray Randall, sterling with simulated turquoise stones; guard chain with filigree details, marked "C.R.R." $175-225.

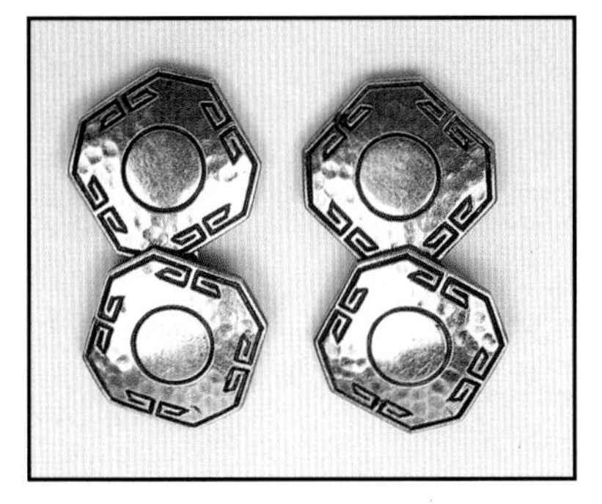

Cufflinks by E.I. Franklin, sterling silver, c. 1920 $75-100.

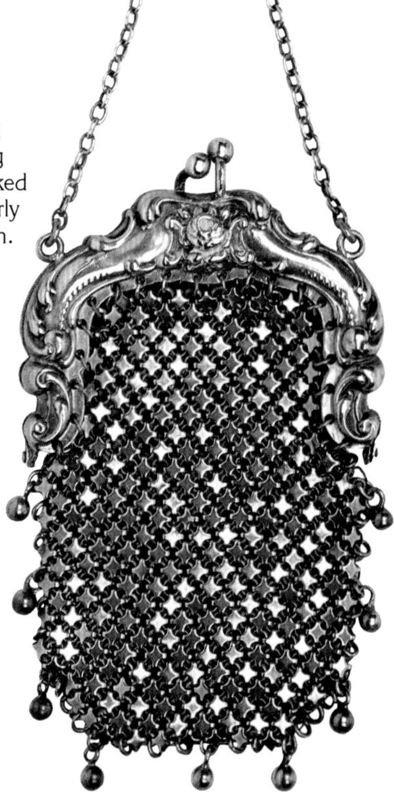

Purse in sterling silver by Whiting and Davis, marked "W & D," an early mark of this firm. $200-295.

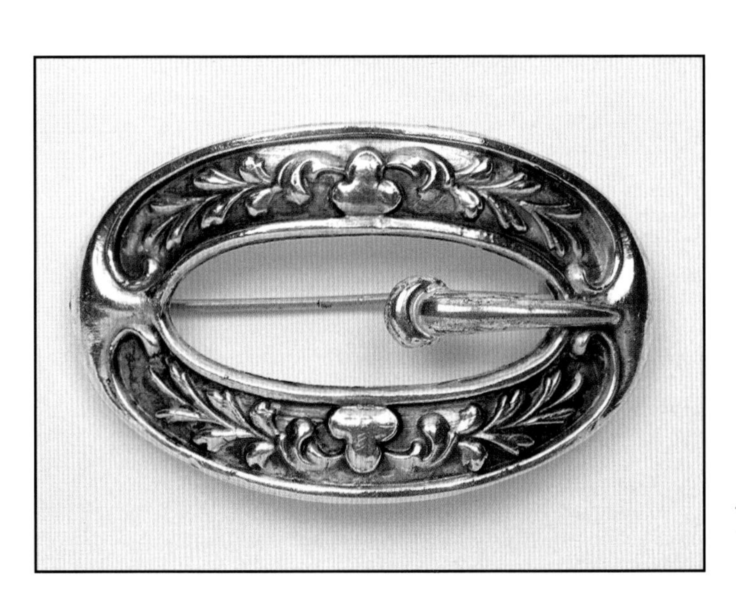

Art Nouveau buckle brooch by the Saart Bros.Co., sterling front, marked "SB" in a globe. $125-150.

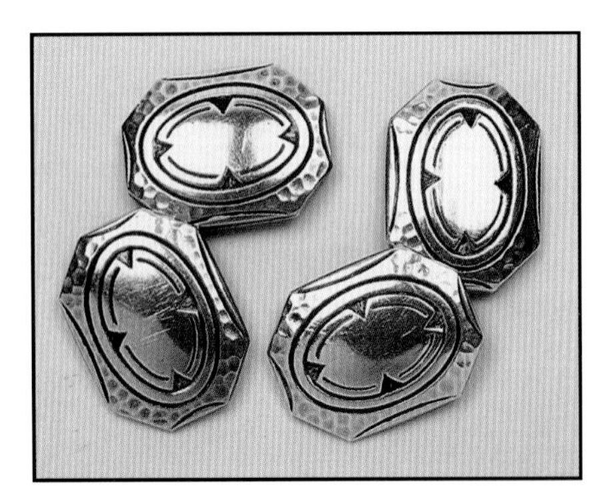

Cufflinks by E.I. Franklin, sterling silver, c. 1920 $75-100.

Belt buckle and fob sets were popular for men c. 1915-1920. This sterling and enamel set in the original box marked "Barrows" for the H.F. Barrows Co., is trademarked "H.F.B." The belt fob slid over the man's belt, with the swivel holding a pocket watch in the trouser pocket. $125-150.

Belt fobs by C.A. Marsh and Charles Thomae , sterling silver, c. 1920s, the fob at left has black enamel. Marked "Marsh" and "Thomae". $75-125.

Art Deco locket by George Webster Co., sterling with engraved design, gold interior, $125-175.

Art Deco locket by Finberg Manufacturing Company, sterling with engraved design, $125-175.

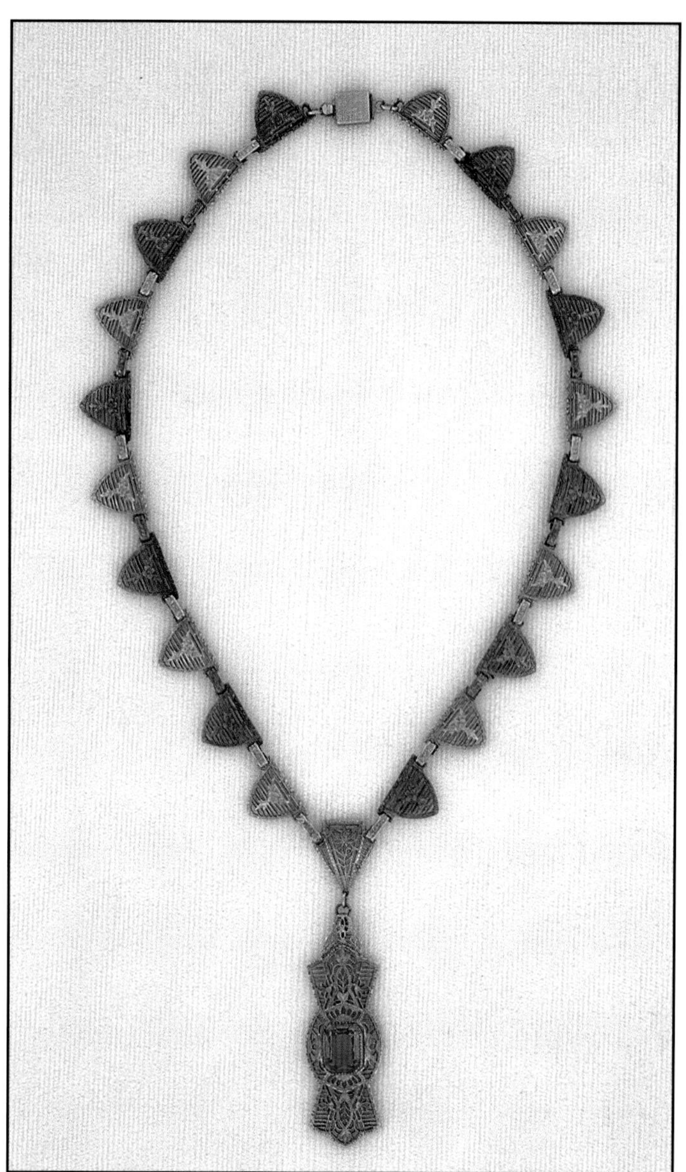

Filigree necklace by Plainville Stock Company. This necklace is a combination of gold and silver metals, showing the level of hand construction in Attleboro costume jewelry of the 1920s. Marked "P.S. Co." Set with faceted glass stone resembling aquamarine. $200-300.

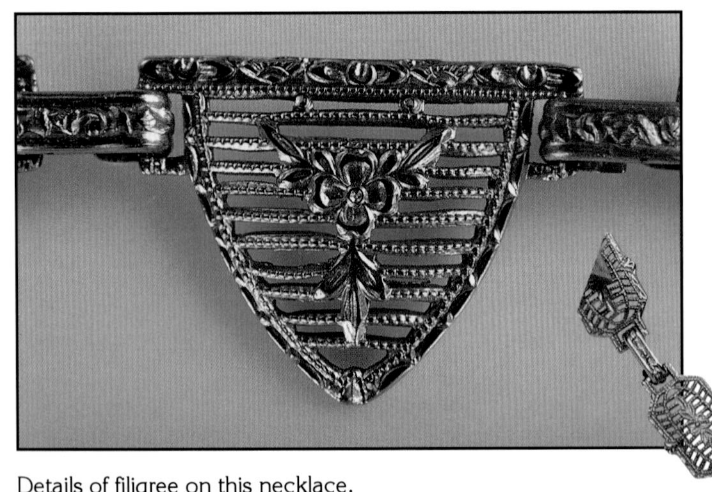

Details of filigree on this necklace.

Filigree bracelet by Plainville Stock Company. The popularity of platinum filigree jewelry c. 1920s prompted three Attleboro jewelry makers, Plainville Stock, A.L. Lindroth, and Ripley and Gowen—R & G LaMode, to offer similar designs as costume jewelry versions. This bracelet is made of a chromium-plated metal, set with a faceted stone, resembling pink sapphire. Marked "P.S. Co." on the clasp. $125-200.

Art Deco straight line bracelets, c. 1925. Top: R. F. Simmons sterling filigree bracelet with simulated sapphires and amethyst stones, marked "Sterling Simmons, Pat. Apld. For". Below: Leach & Miller sterling and rhinestone bracelet with the trademark "L & M in leaves". The fashion for diamond and sapphire line bracelets during the Art Deco period inspired Attleboro jewelers to create quality costume jewelry after costly platinum designs. $150-200.

Locket by Bliss Bros., engraved pattern on sterling silver, c. 1925, $50-75.

Art Deco silhouette pin, marked "Bates & Bacon", gold filled, c. 1925, $125-175.

Art Deco earrings by E.I. Franklin, sterling silver, marcasite, and onyx, c. 1925, all original, measuring 2", $225-300.

Snake bracelet, gold on sterling by Whiting and Davis, marked "W & D." c. 1920s, $350-450.

Chapter 4

Enamels

Peacock brooch by Charles M. Robbins, who traveled to China for six months in 1891 to learn the enamel process. The Robbins Company at one time operated the largest enameling department in the world. This sterling enamel pin was pictured in the 1908 Robbins color enamel jewelry catalog, included in this chapter. $500-750.

Water lily sash ornament by Watson. This Art Nouveau sterling enamel brooch measures 2.75 x 1.5" and is marked with a "crown, W, and lion". The sash ornament, also called a sash pin, was used to fasten a ladies' belt c. 1900. At least ten other Attleboro jewelry makers advertised sash pins in 1910. $450-550.

Calla lily brooch by Watson. Art Nouveau sterling enamel 2.5" x 1". Like many other Watson pieces from the early 1900s, this pin is marked "Genuine Cloisonne" in addition to the "crown, W, and lion" trademark. $350-450.

Art Deco bar pins in sterling enamel. Pin with diagonal green and black lines and pink cabochon stone by E. I. Franklin, marked "E.I.F." and a narrow bar pin with guilloche enamel by G.C. Hudson, with the "G.C.H." trademark, both c. 1925. *Enamel jewelry by these two Attleboro makers is scarce.* $250-450 each.

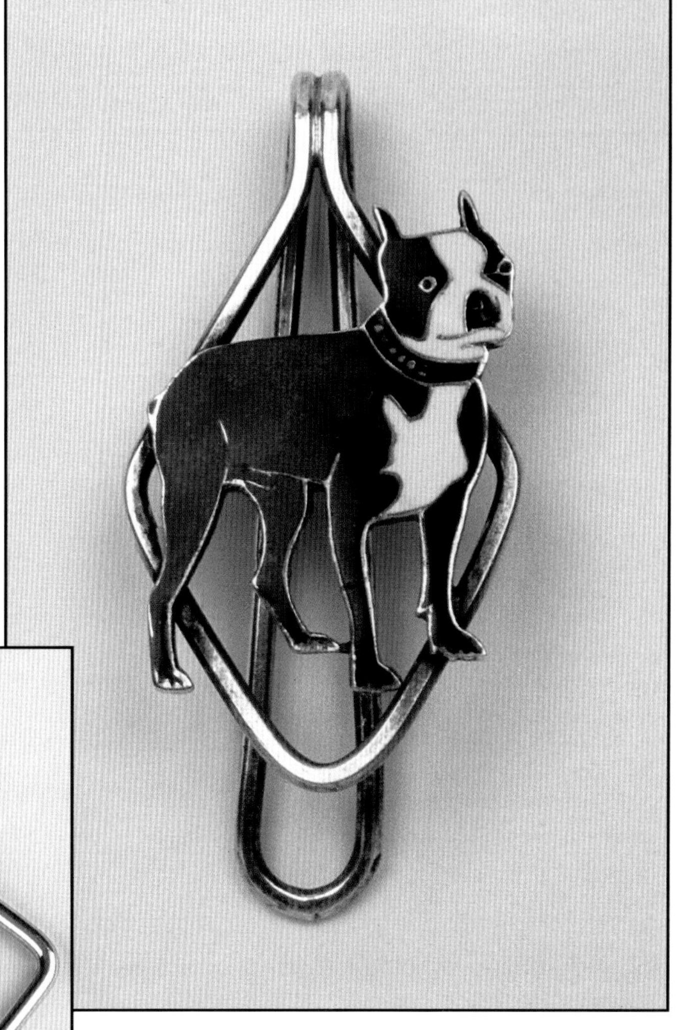

Boston Terrier money clip by Thomae. This company offered a collection of sterling enamel dogs, featuring realistic representations of many different breeds, pins and money clips, c. 1925. $125-225.

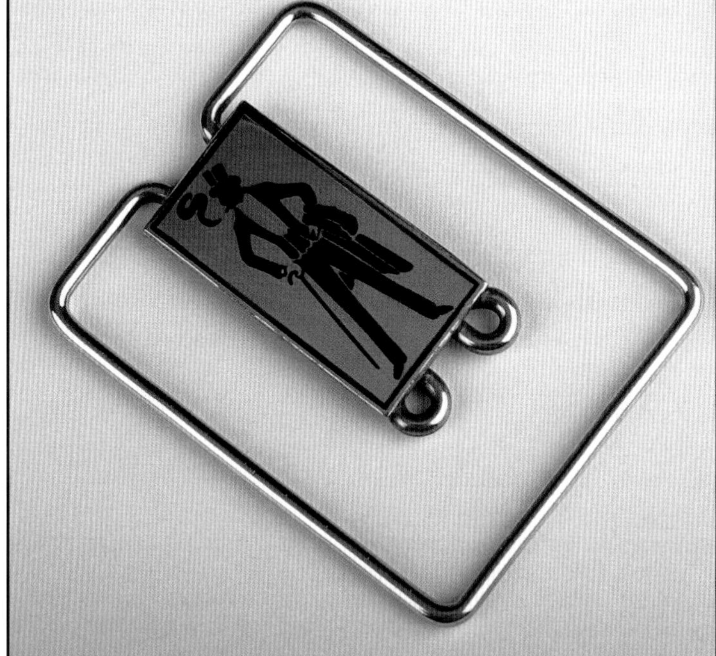

Art Deco money clip by Webster, sterling enamel. The gentleman in top hat and tails was a favorite design c. 1925. $175-225.

Bouquet brooches were popular in the early 1900s, curved to accept a flower or corsage, held in place by a 1/2" pin in the middle. This example is by R. Blackinton in sterling enamel, and shows hand painted roses. $175-250.

Shamrock money clip, Thomae, enamel on sterling silver, marked "Thomae Co., Attleboro, Mass. Sterling", c. 1925. $100-150.

Iris bar pin and oval pin by Charles M. Robbins. Top: Art Nouveau sterling and enamel, 2.8" long, pictured in the 1908 Robbins color catalog. Below: sterling and enamel 2" pin. Both marked "CMR" in a diamond. These pins would have been popular items during the suffragette movement, as women crusading for the right to vote during the early 1900s wore jewelry with "white, purple, and green" $250-350 top, 125-175 below.

Enamel pin by Watson. Like many Attleboro enamel items, this is an example of guilloche, translucent enamel applied over a patterned surface. In the center is a gold bird, called a pallion. Marked "genuine cloisonné" with "crown, W, and lion", 1.75", c. 1920. $200-250.

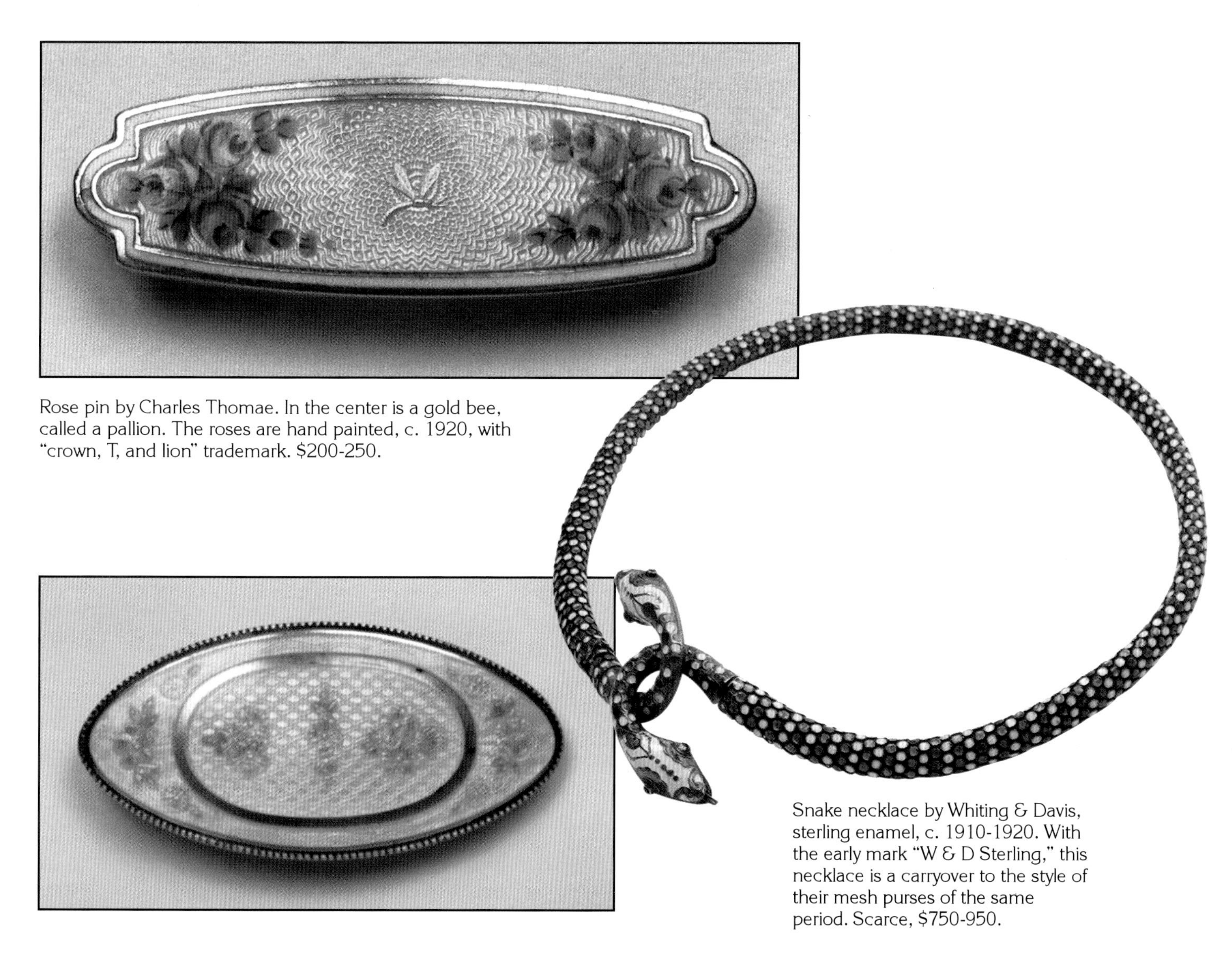

Rose pin by Charles Thomae. In the center is a gold bee, called a pallion. The roses are hand painted, c. 1920, with "crown, T, and lion" trademark. $200-250.

Snake necklace by Whiting & Davis, sterling enamel, c. 1910-1920. With the early mark "W & D Sterling," this necklace is a carryover to the style of their mesh purses of the same period. Scarce, $750-950.

Pink flower pin by Charles M. Robbins. Enamel on sterling, beaded edge, 1.8", marked "champleve sterling" $125-175.

Guilloche rouge or pill box, turquoise enamel on sterling by Webster, vermeil interior with mirror, 1", "W & arrow" trademark, c. 1925. $200-300.

Guilloche pillbox, enamel on sterling, by Charles Thomae, c. 1925, marked with "crown, T, and lion". $225-350.

Guilloche picture locket by R. Blackinton, c.1920. Enamel on sterling, 1.75", opens to hold two photographs. The interior is vermeil, showing this locket's quality. $350-550.

Sweater or cape clips, sterling enamel by Webster. $45-95 each.

Bow pin, sterling enamel with rhinestones by Robbins Co., c. 1925. $100-200.

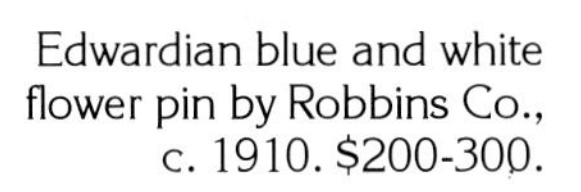

Edwardian blue and white flower pin by Robbins Co., c. 1910. $200-300.

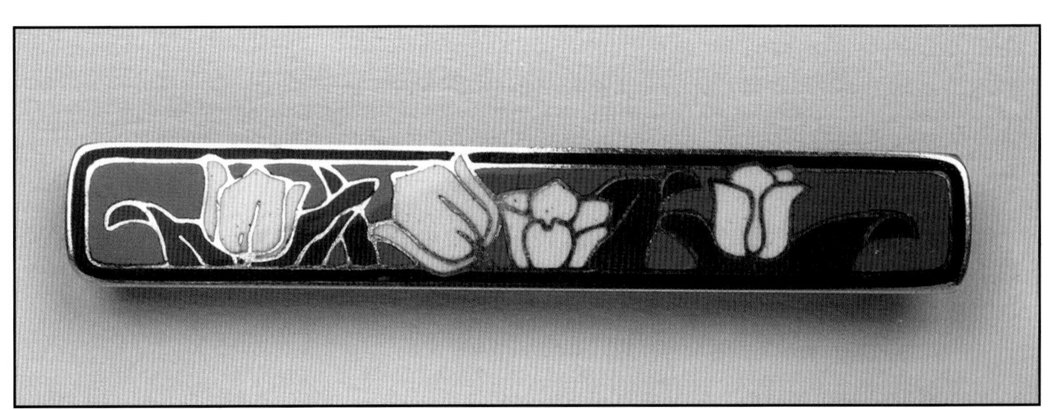

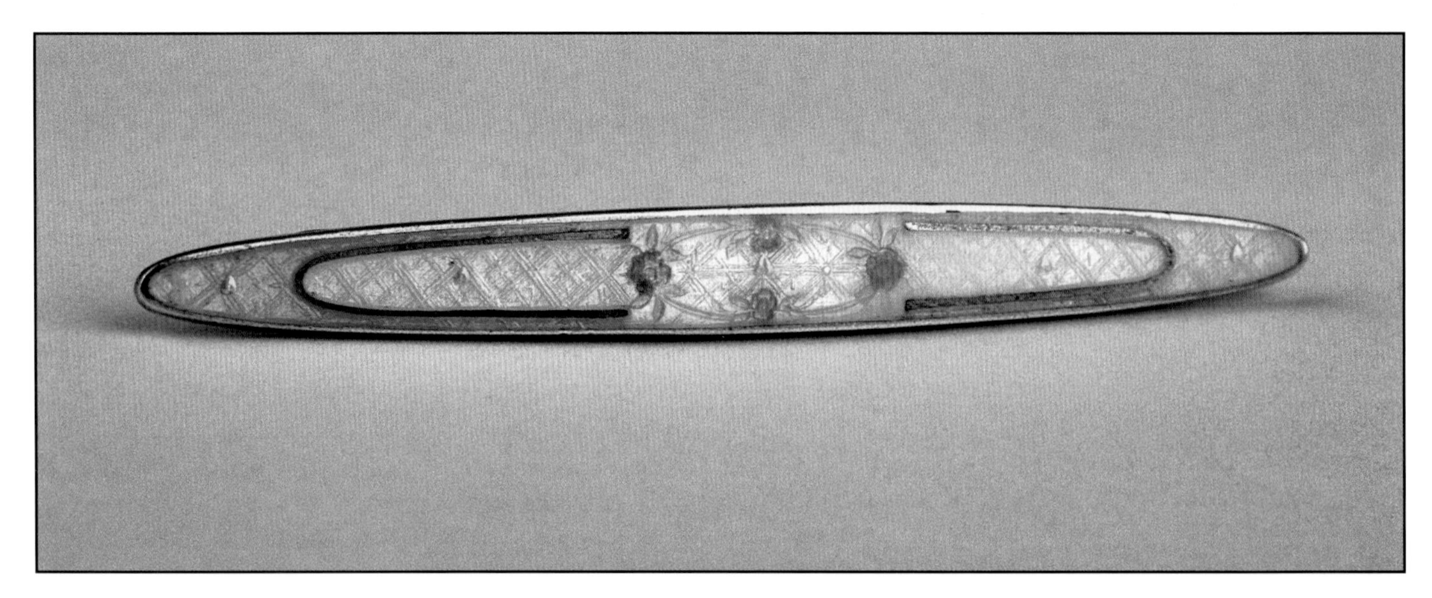

Bar pins in sterling enamel
by Thomae, c. 1920
$150-225 each

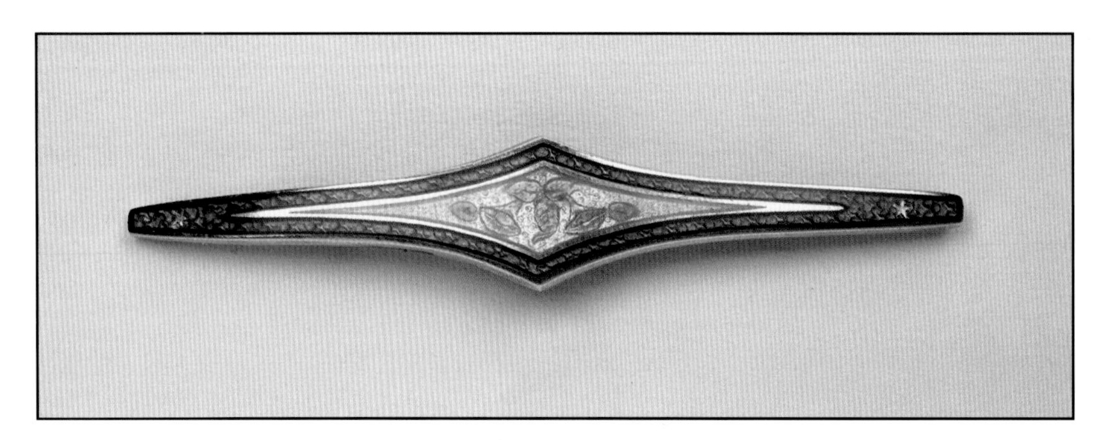

Bar pin in sterling enamel by Watson, c. 1920. $125- 195.

Bar pins in sterling enamel by Watson, c. 1915. $100-225 each.

Tiger lily enamel brooch by Watson, marked "Genuine Cloisonne, Sterling" with crown, W, and lion trademark, 2". $200-300.

White lily brooch by Watson, sterling enamel, 1.3", $100-175.

Cufflinks in sterling enamel; left by Watson and right by Thomae with Masonic emblems. Left: $200-250; right: $100-125.

Princeton pennant and St. Petersburg, Florida pins are examples of college and souvenir jewelry by Robbins Company in the early 1900s. $50-125 each.

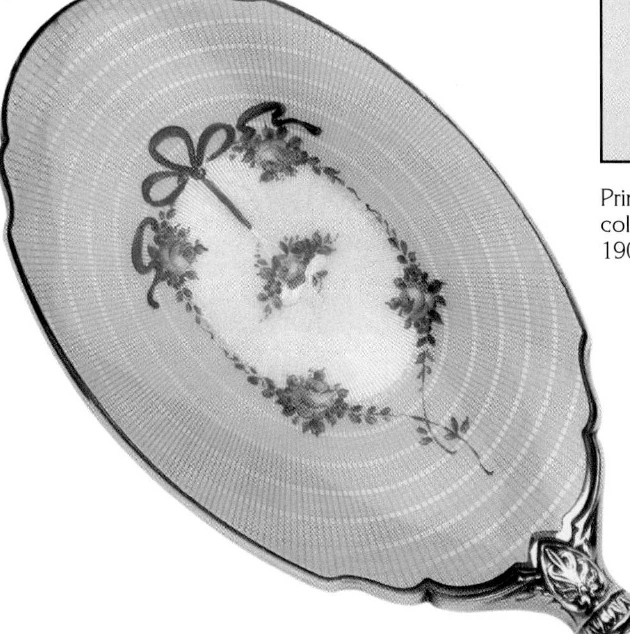

Hand Mirror by Saart Brothers, sterling enamel, c. 1925, $350-500.

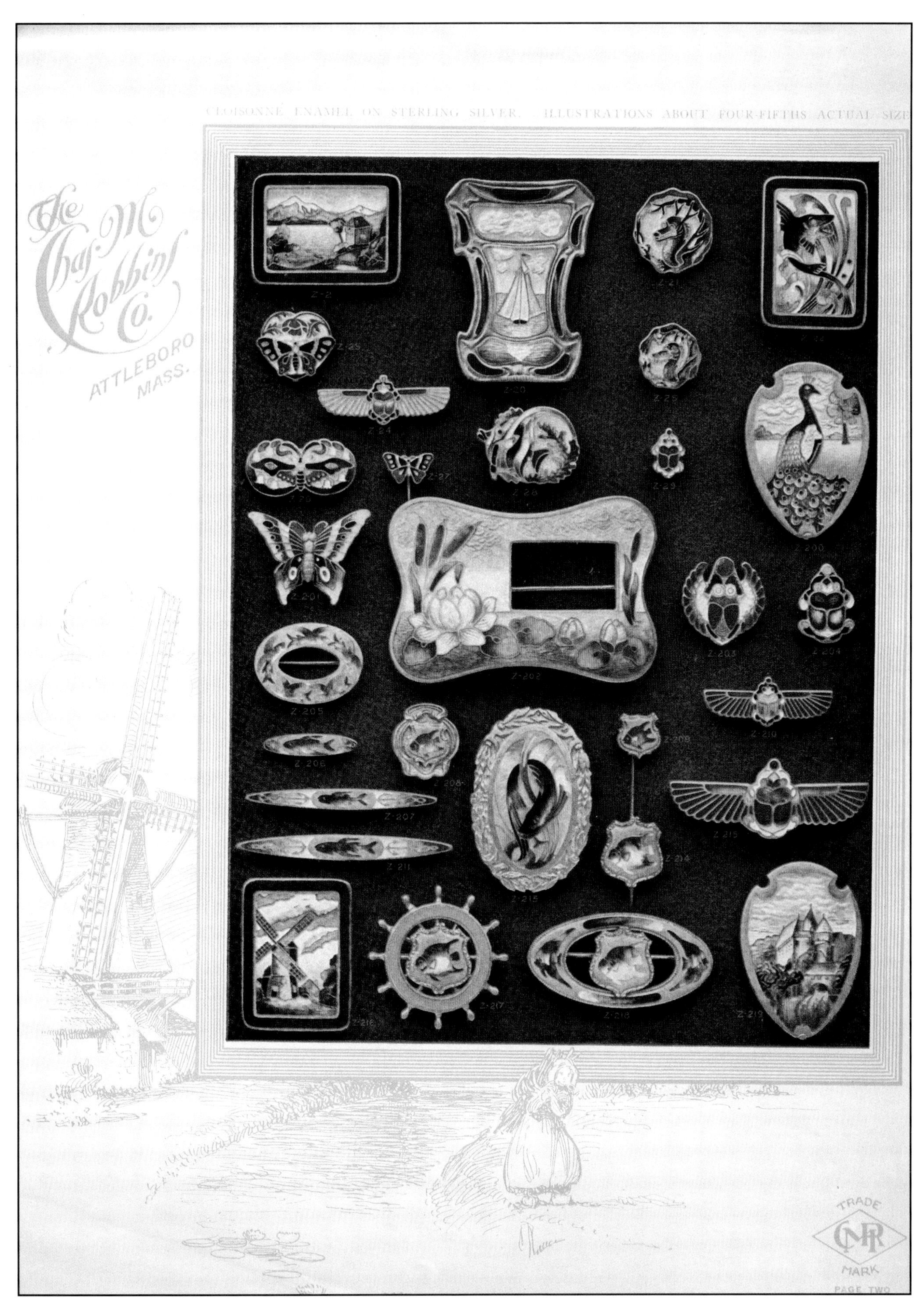

Robbins 1908 Color Enamel Jewelry Catalog

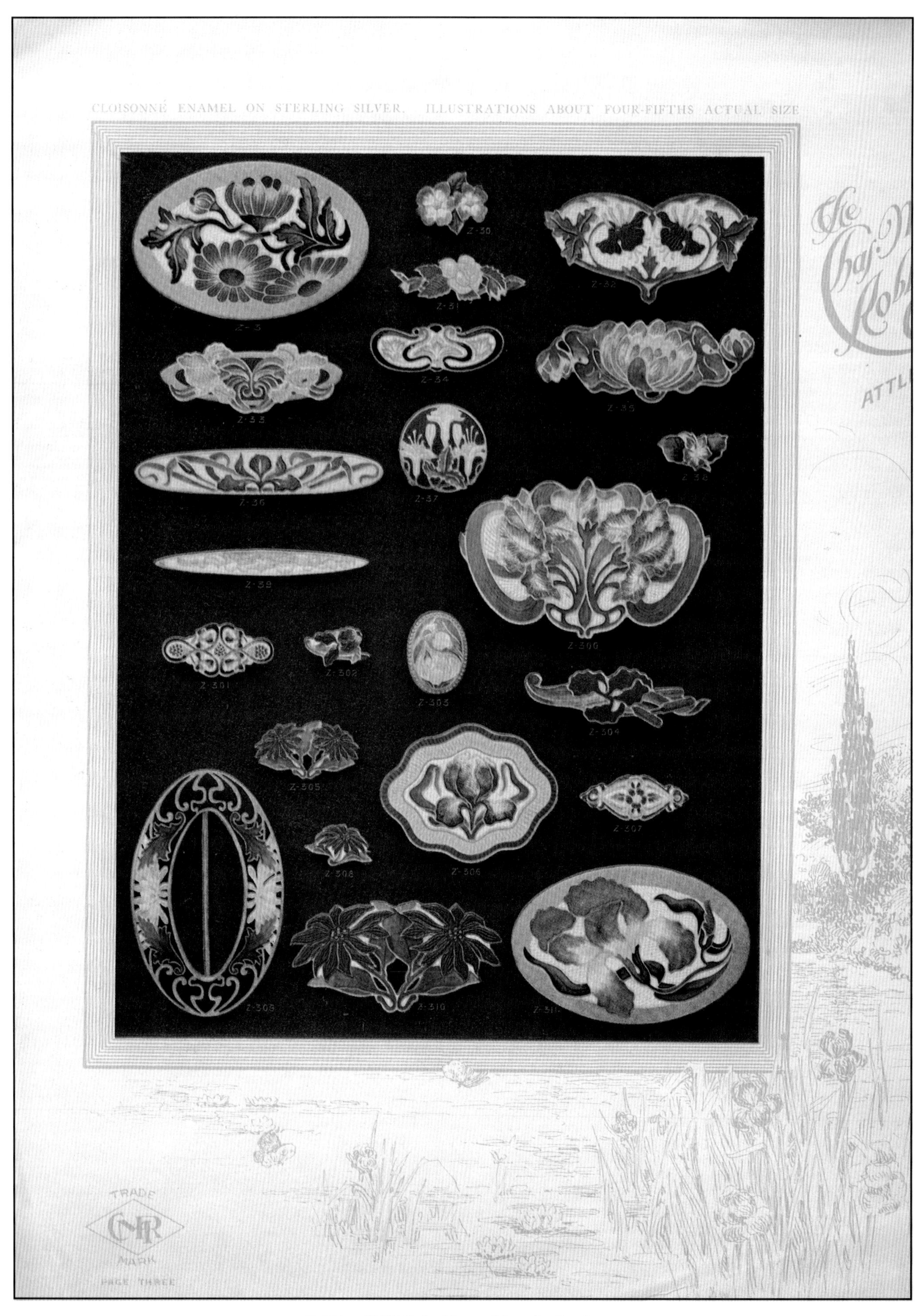

Robbins 1908 Color Enamel Jewelry Catalog

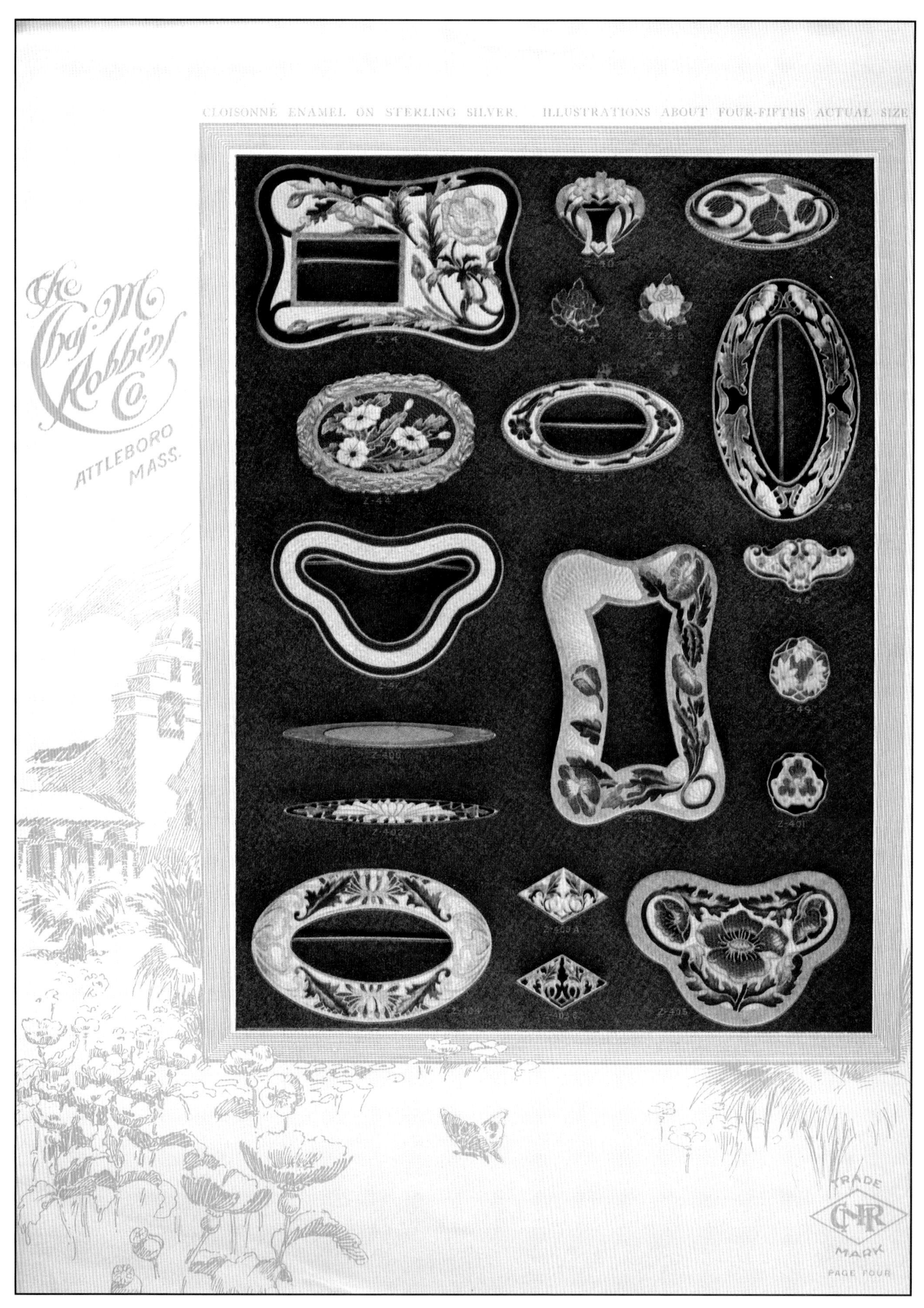

Robbins 1908 Color Enamel Jewelry Catalog

Robbins 1908 Color Enamel Jewelry Catalog

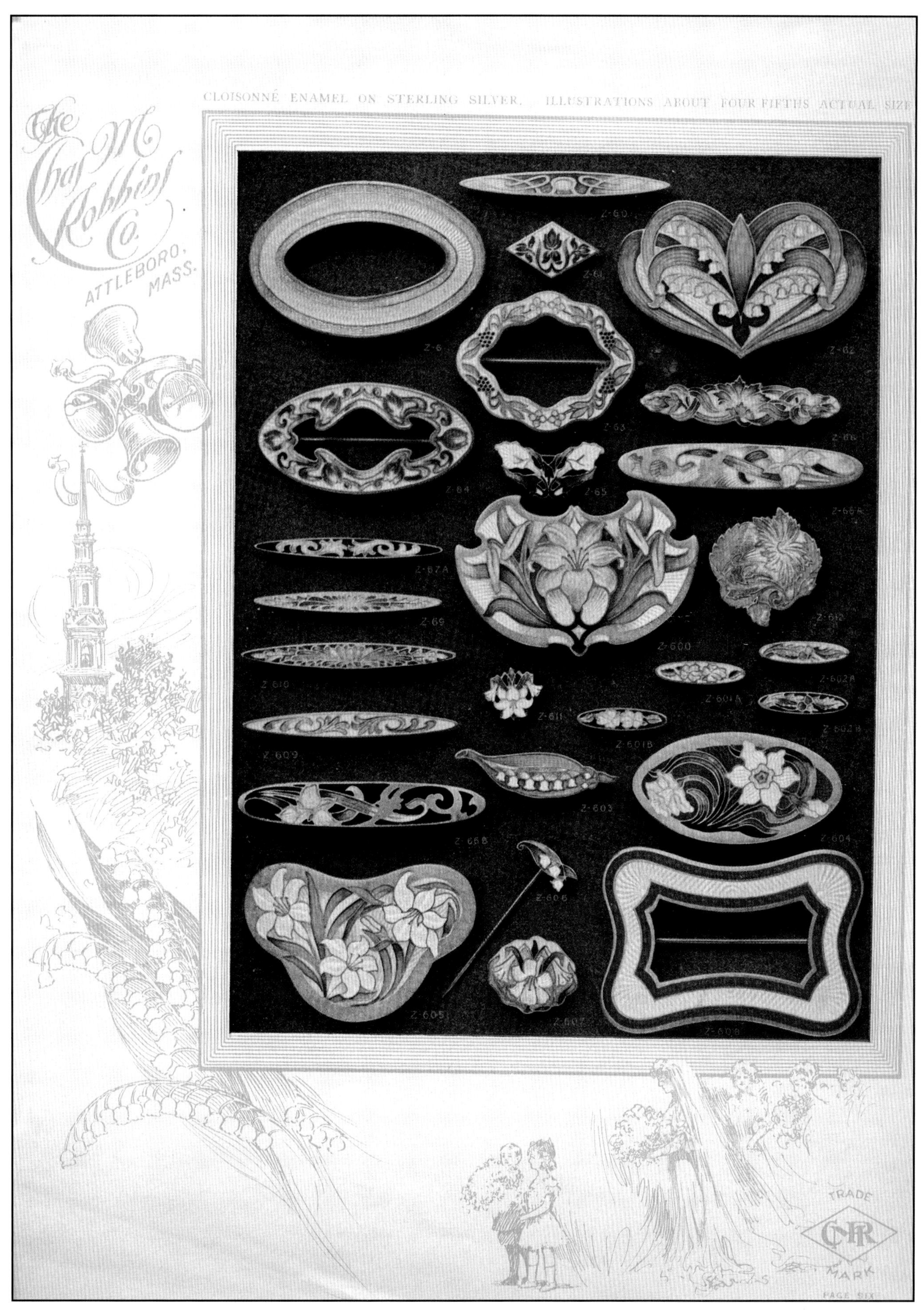

Robbins 1908 Color Enamel Jewelry Catalog

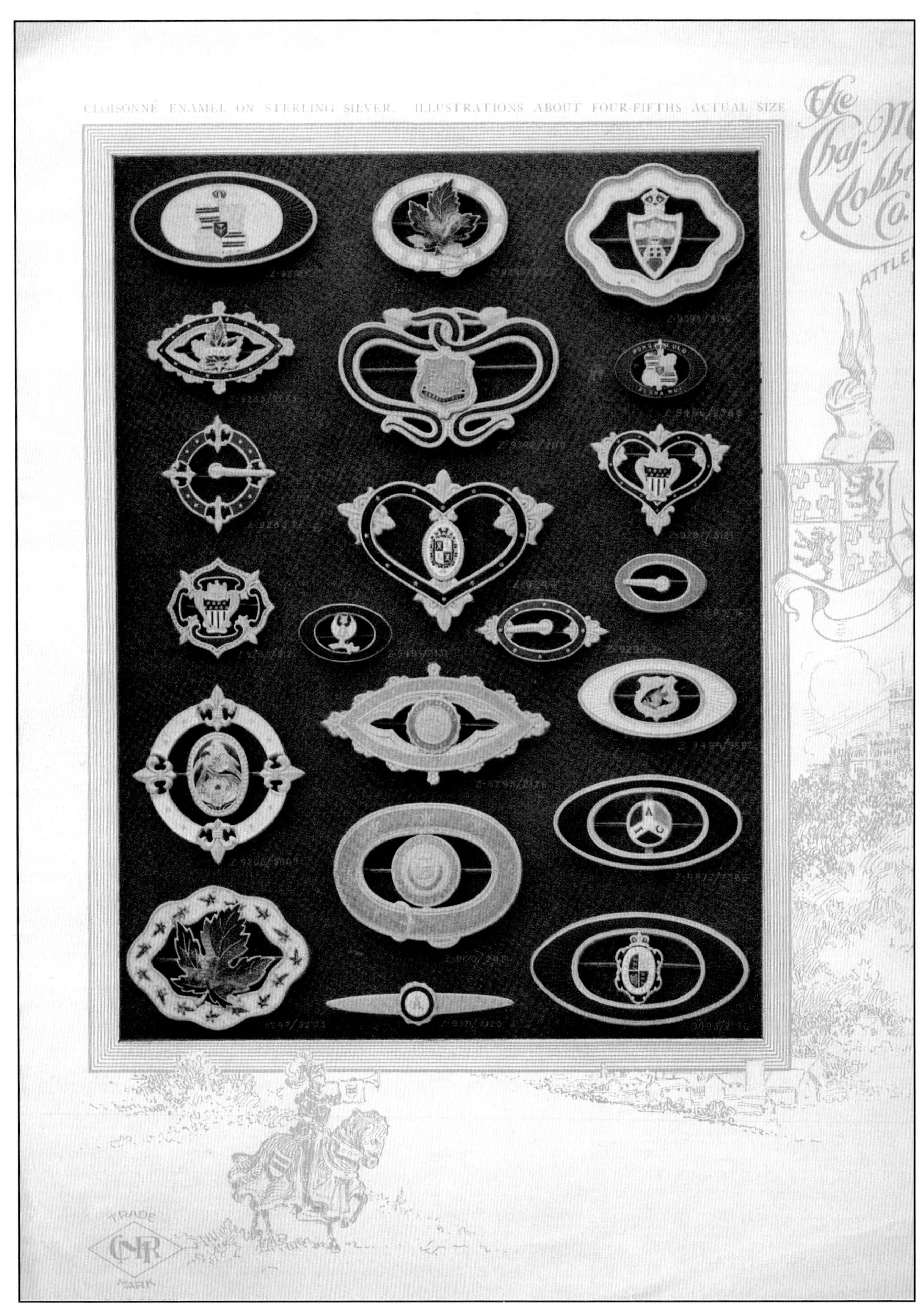

Robbins 1908 Color Enamel Jewelry Catalog

Chapter 5

Paye and Baker Company

Art Nouveau lady brooch by Paye & Baker, sterling silver, c. 1910, marked "P & B" in three hearts. $400-500.

Bluebird brooches by Paye & Baker, sterling enamel, 3" and 1.5", marked "P & B" in three hearts, $125-225 each.

Billiken pin by Paye & Baker, sterling silver, c. 1910-1915, marked "P & B" in three hearts and with Billiken trademark. During his popularity, the Billiken was considered to be good luck and "the god of things as they ought to be". $100-150.

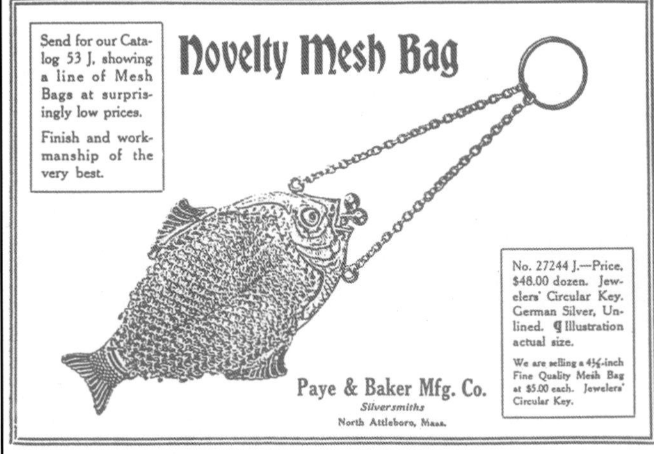

Paye and Baker Bears, "Billiken," and Fish purse ads"

Chapter 6
Match Safes

Art Nouveau lady, James E. Blake Co., c. 1900, back stamped "SterlinE". $65-85.

Art Nouveau lady & flowers, James E. Blake Co., c. 1900, back stamped "SterlinE". $65-85.

Floral design, James E. Blake Co., c. 1900, back stamped "SterlinE". $45-65.

Lady in boat, James E. Blake Co, c. 1900, back stamped "SterlinE". $100-125.

Figural bottle or flask, James E. Blake Co., c. 1905; leather wrapped, back stamped "SterlinE". $75-95.

Art Nouveau lady smoking and drinking, James E. Blake Co., c. 1900; silver plated, also made in sterling and SterlinE. $100-125.

Fishing motif, James E. Blake Co., c. 1900; back stamped "SterlinE", also made in sterling. $100-125.

Floral motif, James E. Blake Co., c. 1900, bezel stamped "SterlinE". $40-50.

Art Nouveau nude, James E. Blake Co., c. 1900, back stamped "SterlinE". $75-95.

Lady in the garden, James E. Blake Co., c. 1900; left, bezel stamped "SterlinE"; right, bezel stamped "Sterling", gold wash inside. Left $75-100. Right $125-150.

Vertical line design, James E. Blake Co., c. 1910; bezel stamped "Sterling", catalog #670 and Blake's maker mark. Gold wash inside. $75-100.

Art Deco styling, James E. Blake Co., c. 1910, bezel stamped "Sterling" plus Blake's maker mark, gold wash inside. $75-100.

Rococo motif, F. S. Gilbert, c. 1895; bezel stamped "Sterling G", gold wash inside. $ 75-95.

Art Nouveau lady & flowers, F. S. Gilbert, c. 1895, bezel stamped "Sterling G". $90-110.

Floral motif, F.S. Gilbert, c. 1900, bezel stamped "Sterling G". $75-95.

Floral motif, F. S. Gilbert, c. 1900, bezel stamped "Sterling G" $65-85.

Floral & vine motif, F. S. Gilbert, c. 1900, bezel stamped "Sterling G". $75-95.

Floral motif, F. S. Gilbert, c. 1900, bezel stamped "Sterling G". $65-85.

Rose motif, F. S. Gilbert, c. 1900, bezel stamped "Sterling G". $65-85.

Art Nouveau lady, F. S. Gilbert, c. 1900, bezel stamped "Sterling G". $75-95.

Three ladies with flowers, F. S. Gilbert, c. 1900, bezel stamped "Sterling G", gold wash inside. $275-300.

Enameled coat-of-arms, F. S. Gilbert, c. 1900, bezel stamped "Sterling G". $100-125.

Serpent motif, F. S. Gilbert, c. 1900, bezel stamped "Sterling G", shown with shiny and satin finishes. $125-150.

Brotherhood Protective Order of Elks motif, F. S. Gilbert, c. 1900, bezel stamped "Sterling G", gold wash inside. $175-200.

Serpent motif, F. S. Gilbert, c. 1900, bezel stamped "Sterling G". $125-150.

Bowling motif, F. S. Gilbert, c. 1895, bezel stamped "Sterling G", gold wash inside. $375-395.

Hidden photo, F. S. Gilbert, c. 1895, bezel stamped "Sterling G". Hidden photo release inside. $250-275.

American Indian, F. S. Gilbert, c. 1895, bezel stamped "Sterling G", gold wash inside. $450-500.

Rose motif, F. S. Gilbert, c. 1895, bezel stamped "Sterling G", gold wash inside. $225-250.

Hinged door, Webster Company, c. 1900, bezel stamped "Sterling" and Webster's mark, gold wash inside. $275-300.

Floral motif, Webster Company, c. 1900, bezel stamped "Sterling" and Webster's mark, gold wash inside. $85-100.

American Indian chief, Webster Company, c. 1900, bezel stamped "Sterling" and Webster's mark, gold wash inside. $700-800.

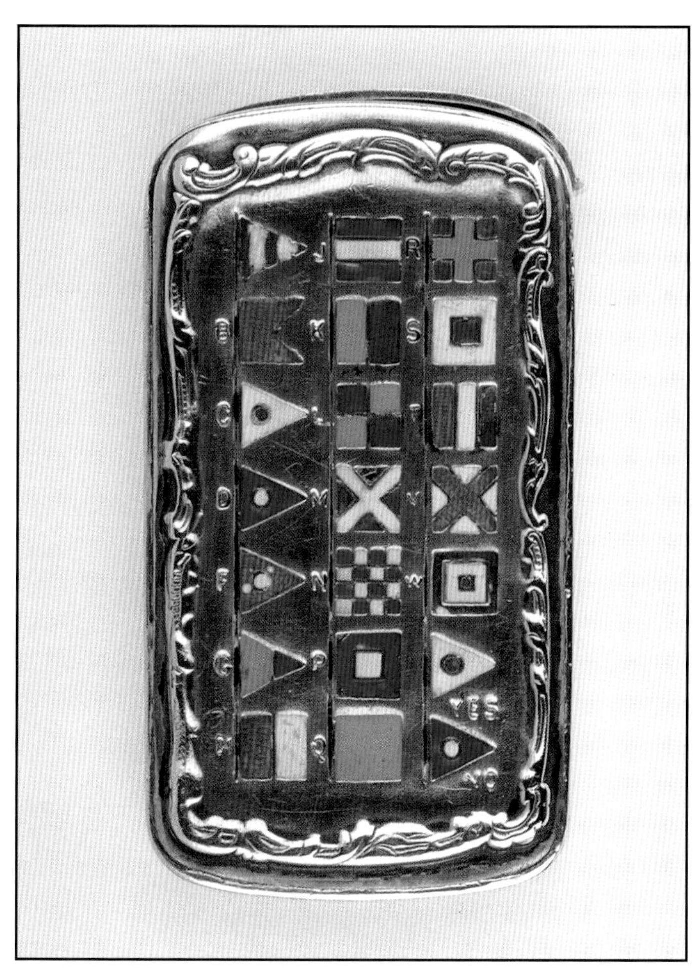

Enamel signal flags, Webster Company, c. 1900, inside lid stamped "925/Sterling" and "Webster Company". Back mother-of-pearl. $425-475.

Art Nouveau nude on mother-of pearl, Webster Company, c. 1900, inside lid stamped "925/Sterling" and "Webster Company", mother-of-pearl panels on front and back. $350-395.

Vertical engraved lines, Webster Company, c. 1910, bezel stamped "Sterling" and Webster's mark, gold wash inside. $85-100.

Fox chase scene, Bristol Mfg. Co., c. 1900, Bristol Silver, bezel stamped "Bristol Silver". Examples are also made in sterling. $65-85.

Art Nouveau standing nude, Bristol Mfg. Co., c. 1900, Silveroin, bezel stamped "Silveroin". $60-75.

Floral motif, Bristol Mfg, Co., c. 1900, Silveroin, bezel stamped "Silveroin". $40-50.

Nude riding horses, Bristol Mfg. Co., c. 1900, Silveroin, bezel stamped "Silveroin". $100-125.

Fishing scene, Bristol Mfg. Co., Silveroin, bezel stamped "Silveroin". $100-125.

Figural barrel and figural barrel with lady smoking, both Bristol Mfg. Co., c. 1900; at left, Silveroin, bezel stamped "Silveroin"; at right, bezel stamped "Sterling", gold wash inside. $90-110 Left, $275-300 Right.

Plain, Frank M. Whiting, c. 1910, bezel stamped "Sterling, catalog # 207 and Whiting's mark. $65-85.

Slant lid with engraved floral motif, Frank M. Whiting, c. 1900, bezel stamped "Sterling" catalog # 81 and Whiting's mark. $75-90.

Couple in seashell, Frank M. Whiting, c. 1893, bezel stamped "Sterling" and Patent June 6, '93. Cigar cutter built into lid. Front and back illustrated. $350-400.

Nude standing on swan's back, Frank M. Whiting, c. 1893, bezel stamped "Sterling" and Patent June 6, '93. Cigar cutter built into lid. $350-400.

Scalloped edges, Frank M. Whiting, c. 1895, bezel stamped "Sterling 925/1000" and Whiting's mark. $85-100.

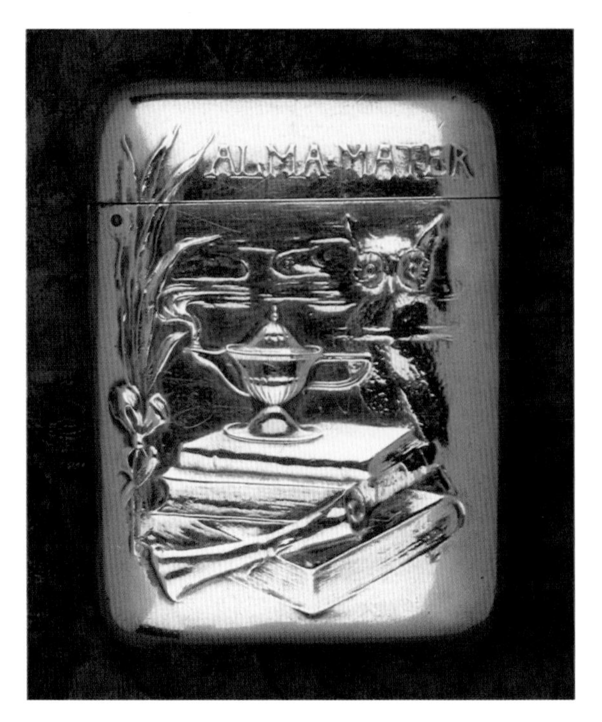

College-alma mater motif, R. Blackinton & Co., c. 1900, bezel stamped "Sterling" catalog #1302 and Blackinton's mark, gold wash inside. Front and back illustrated. $1300-1500.

Speeding auto, R. Blackinton & Co., c. 1905, bezel stamped "Sterling", catalog #1332, and Blackinton's mark, gold wash inside. $500-550.

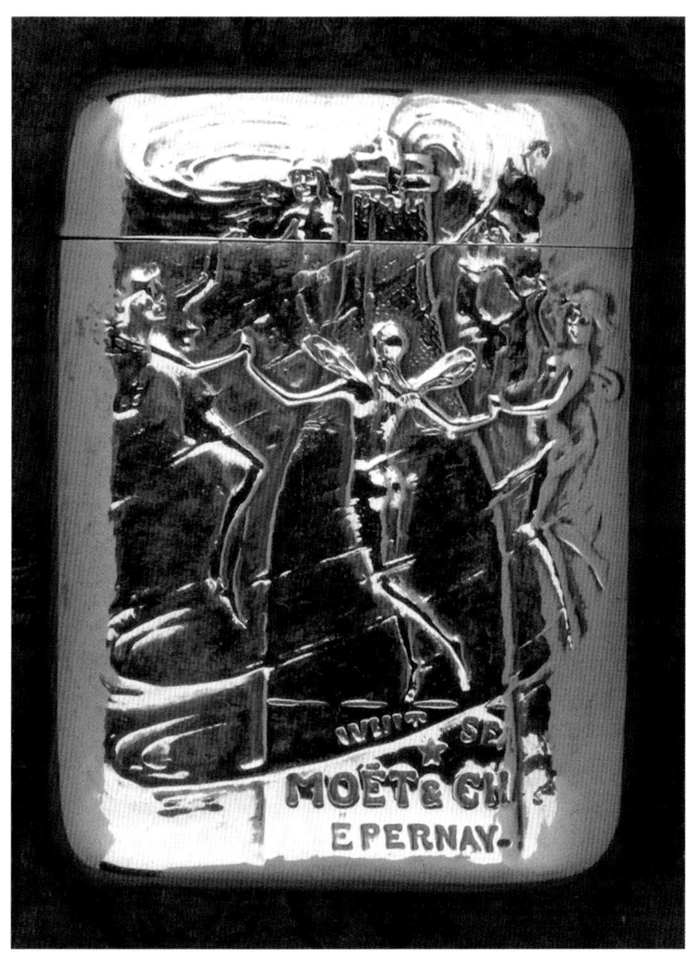

Moet champagne, R. Blackinton & Co., c. 1905, bezel stamped "Sterling" catalog #1708 and Blackinton's mark, gold wash inside. $750-900.

Sterling and copper, R. Blackinton & Co., c. 1900, bezel stamped "Sterling & Copper" catalog # 359, and Blackinton's mark. $450-500.

Sterling & copper, R. Blackinton & Co., c. 1900, bezel stamped "Sterling & Copper", other marks worn. $400-450.

Fish nouveau motif, R. Blackinton & Co., c. 1900, bezel stamped "Sterling" catalog #374, and Blackinton's mark, gold wash inside. $750-950.

Floral surrounded cartouche, R. Blackinton & Co., c. 1900; ruby thumbpiece, back concave and stamped "Sterling" and Blackinton's mark, gold wash inside. $150-200.

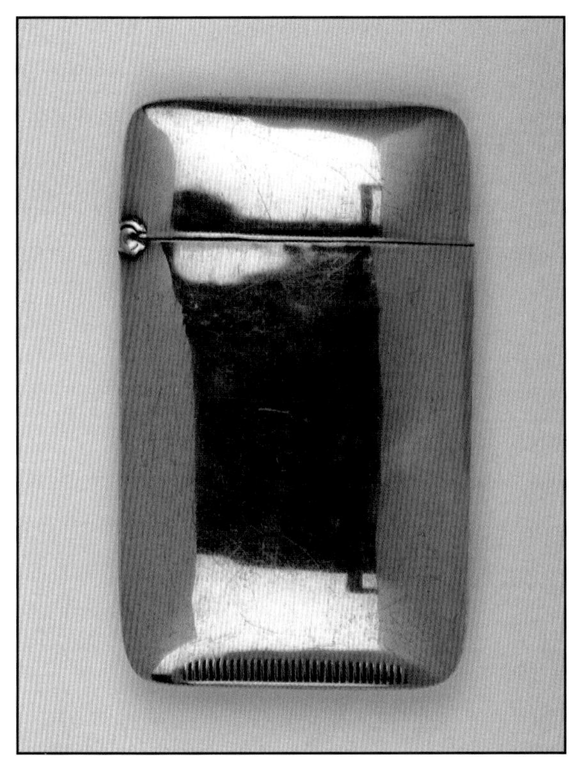

Sterling with gold finish, G. H. French & Co., c. 1900, bezel stamped "Sterling" and French's mark, gold wash inside. $95-110.

Floral motif, Attleboro Chain Co., c. 1900, bezel stamped sterling and Attleboro's mark, gold wash inside. $100-125.

Rose motif, Palmer & Peckham, c. 1895, bezel stamped "Sterling" and Palmer & Peckham's mark. $100-125.

Sleek looking plain design, The Thomae Co., c. 1910, inside stamped "Sterling", catalog #T22 and Thomae's mark, gold wash inside, reverse concave. $100-125.

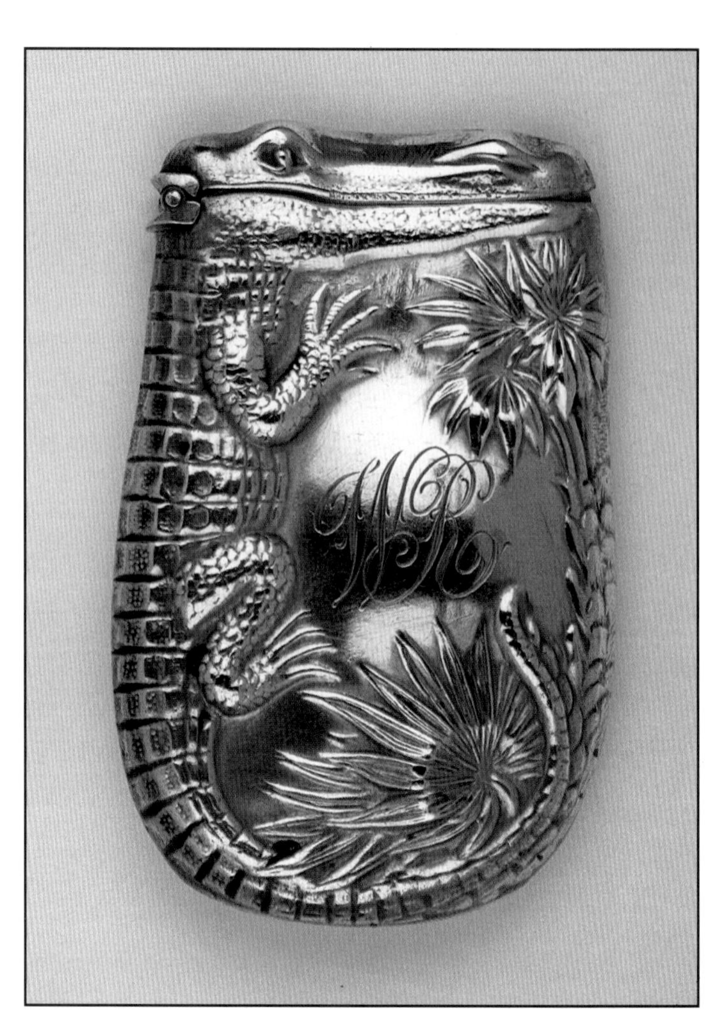

Alligator, Watson Company, c. 1895, bezel stamped "Sterling", catalog # 16 and Watson's mark, gold wash inside. $600-700.

Art Nouveau nude, Watson Company, c. 1900, bezel stamped "Sterling" and Watson's mark, gold wash inside. $125-150.

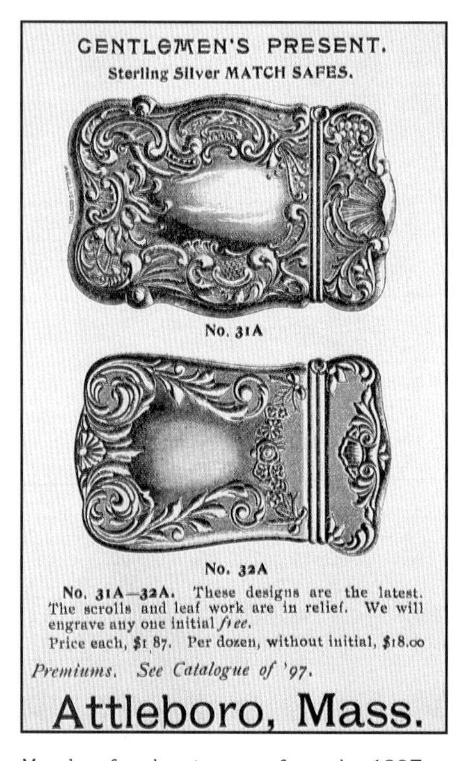

Match safe advertisement from the 1897 McRae and Keeler catalog.

Made by

THE THOMAE CO.

67 Mechanics Street

Attleboro, Massachusetts

NEW YORK	CHICAGO	LOS ANGELES
347 Fifth Ave.	37 So. Wabash Ave.	643 So. Olive St.

Smart Ash Trays and Match Boxes

for SMART PEOPLE ......

Hammered. Solid Sterling Silver

Ash Tray No. 4084. Hammered . $2.10 each, List
Match Box No. 4083. Hammered . 1.80 each, List

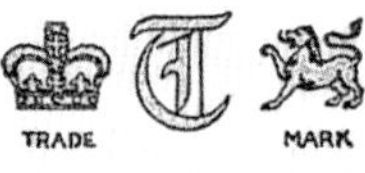

THE THOMAE CO.

ATTLEBORO, MASS.

NEW YORK	CHICAGO	LOS ANGELES
347 Fifth Ave.	37 So. Wabash Ave.	643 So. Olive St.

Advertisement from a Thomae brochure.

Chapter 7

Compacts, Cases, Lighters, and Novelties

Cigarette case by James E. Blake, who was granted a patent to manufacture sterling silver and 14K gold items on January 31, 1905. This case is yellow gold and rose gold on sterling silver, c. 1920, marked "Sterling 14K" with trademark "J.E.B. Co". $400-550.

Napkin holder by James E. Blake, sterling silver and 14K gold, shows the prominence of two-tone metals in the early 1900s. $100-150.

Belt Buckle by Daggett & Clap, sterling silver and 14K gold, c. 1925, marked "D & C", with finely engraved pattern $125-195.

Evans checkerboard enamel lift arm lighter, early 1920s, in mint condition $750-1,000.

Art Deco cigarette case and cigarette case / lighter combination, c. 1925, by Evans. $100-125 each.

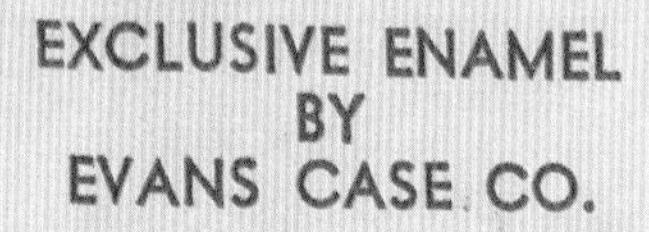

EXCLUSIVE ENAMEL
BY
EVANS CASE CO.

After many years of research, our Laboratory has developed the exclusive enamel process which represents the finish on this product.

Enamel has been used throughout the ages and, in the art world, is associated often with those never-to-be-forgotten artists such as Cellini, Fabergé and Leonardo da Vinci who stated "*Enamel is the eternal art! Enamel is the most lasting, most beautiful and most enduring form of decoration.*"

Enamel is a vitreous glaze fired at 1475° Fahrenheit on a metallic surface and the result is a smooth, glossy, shimmering surface which may be cleaned easily with a damp cloth.

With proper care, we GUARANTEE that our products will last indefinitely.

EVANS CASE CO.
Plainville, Massachusetts

CP12207

"The Evans Atarmist," Art Deco perfume atomizer by Evans Case Co., is blue, black, and silver metal and was found in new condition with instructions in the original box. When the dial on top is turned, the perfumer may be operated. "Exclusive Enamel by Evans Case Co." page reads "After many years of research our Laboratory has developed the exclusive enamel process which represents the finish of this product. Enamel is a vitreous glaze fired at 1475 degrees Fahrenheit on a metallic surface and the result is a smooth, glossy, shimmering surface which may be cleaned easily with a damp cloth. With proper care, we guarantee our products will last indefinitely." Perfumer in mint condition, new in box with instructions, $750-850.

Ladies Silhouette Compact by James E. Blake, sterling and enamel, c. 1920. Two separate compartments with gold vermeil interiors, top and bottom, hold pressed powder and powder rouge in this compact. $350-550.

Art Deco guilloche enamel compact by Evans with hand painted rose decoration. The interior contains both pressed powder and rouge compartments, the latter of which has the original rouge, stamped with the Evans logo. Many Attleboro compact makers provided refills for pressed powder and rouge by mail order. $100-150.

Guilloche enamel compact with hand painted enamel bluebird by Finberg Manufacturing Co., c. 1920s. The compact opens to hold pressed powder, rouge, and a mirror. $350-500.

Enamel compact by Evans, bouquet of flowers set with marcasites, early 1930s, holds pressed powder, rouge, and a mirror. $300-450.

Powder and lipstick case, hand painted, guilloche enamel on goldtone metal, marked Evans. $100-150.

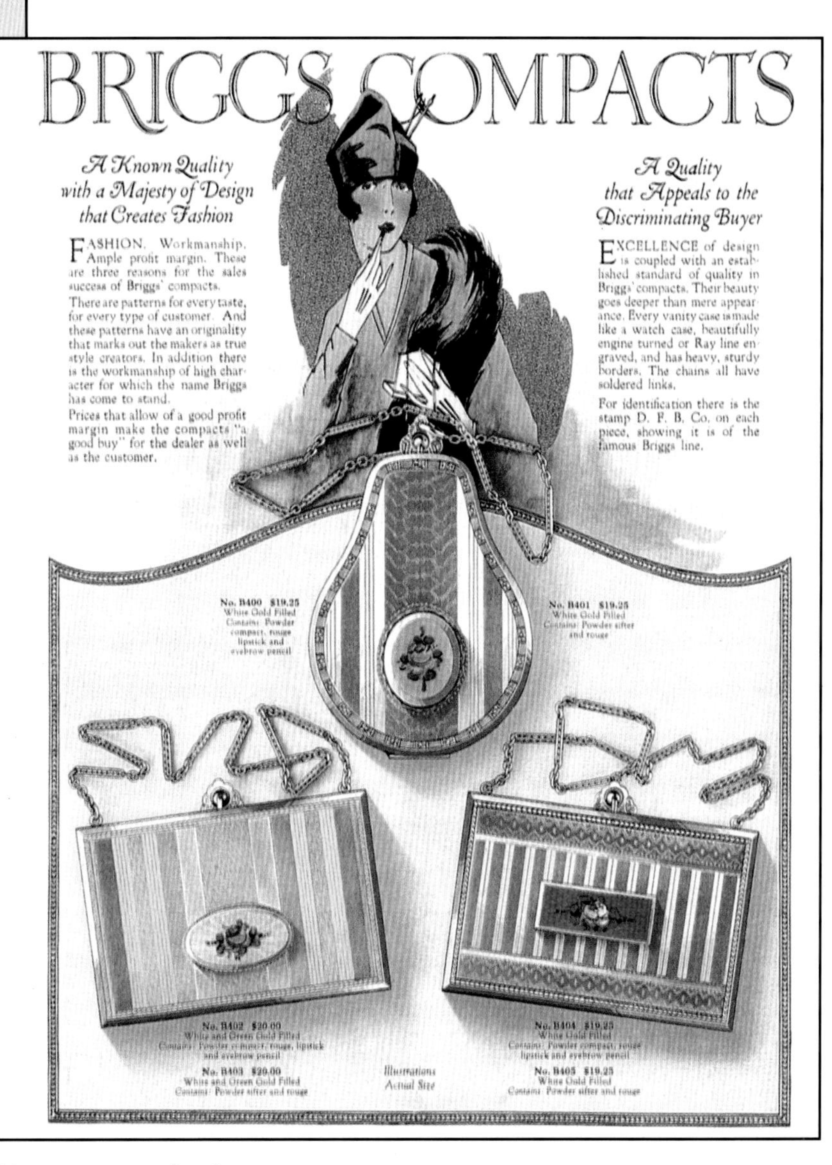

Briggs compact advertisements

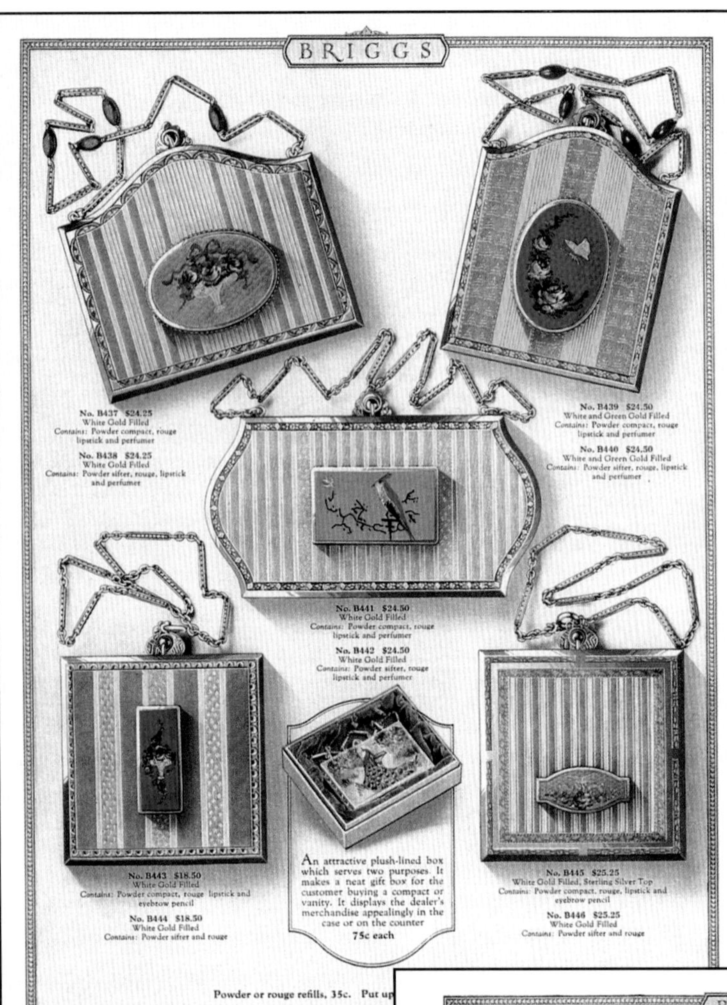

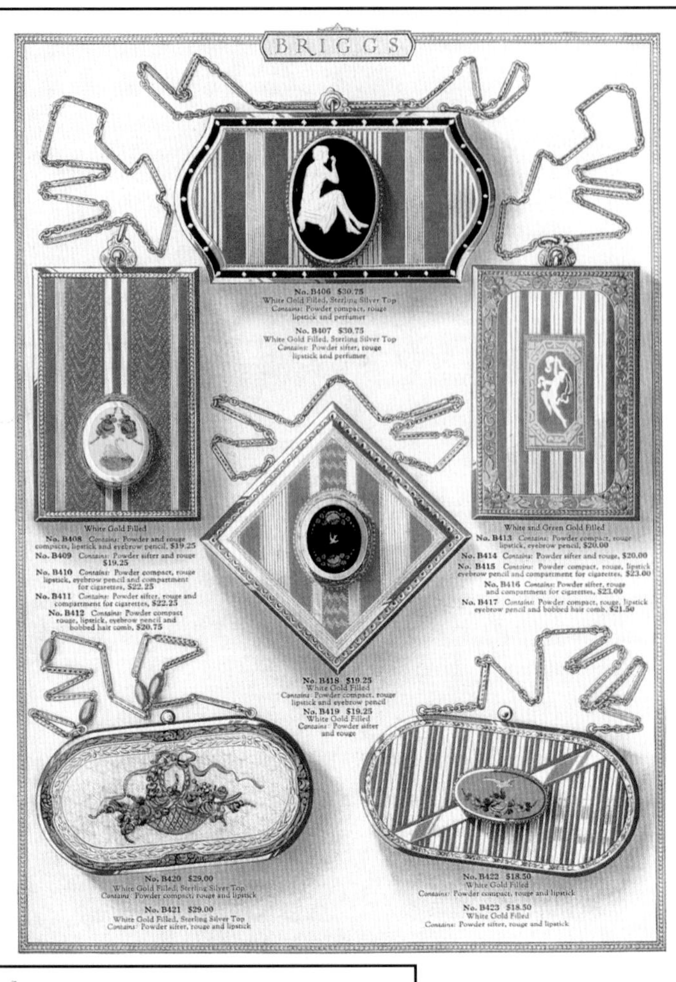

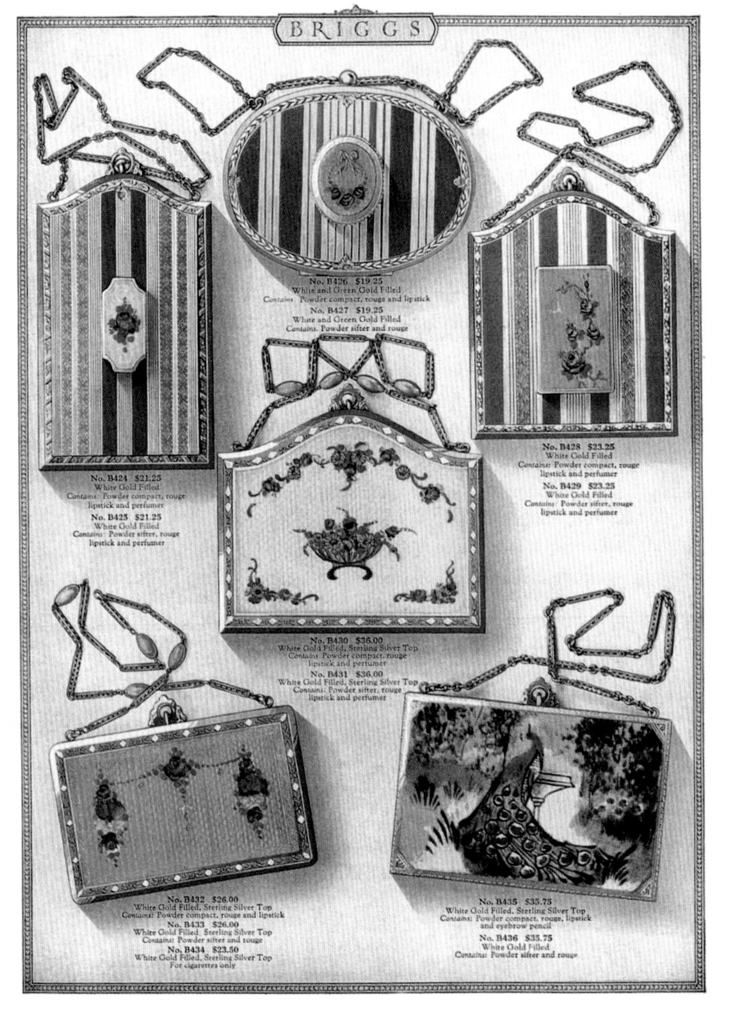

Briggs compact advertisements

Chapter 8

Snap Links: Favorite Cuff Links from the Roaring Twenties

Treasure Chest Gift Box accompanied certain snap links by Baer & Wilde for Kum-a-part in the 1920s. $35-45.

Victorian two-part gold plated cufflinks, late 1800s with a locking mechanism; precursors to the snap links of the 1920s. $65-75 each pair.

StaLokt cufflinks were J.F. Sturdy's Sons' version of the snap link, with a button "to press" and open the cufflink, gold filled. $75-95.

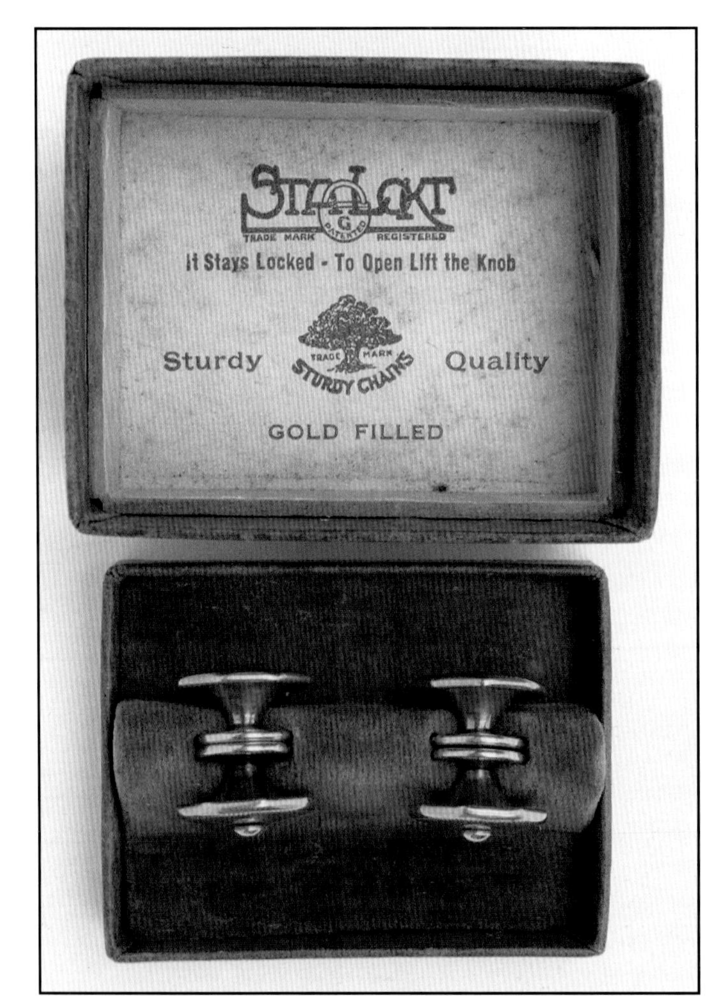

StaLokt cufflinks in their original box with advertising, c. 1925, gold filled. $100-125.

Art Deco enamel snap links, Kum-a-part by Baer & Wilde Co., by far the largest manufacturer of this style of cufflink. $100-125 each.

Art Deco enamel snap links, Kum-a-part by Baer & Wilde. These cufflinks were clearly marked by the maker, often with a patent date of 1923. $75-125.

Art Deco enamel snap links, c. 1925. Every color of the rainbow is found on examples of these enamel cufflinks. $75-125.

Art Deco snap links with high style geometric designs marked "Bliss Bros." are scarce; shown with Kum-a-part cufflinks of the same era. Bliss $200-250, Kum-a-part $50-75.

Kum-a-part snap links with glass stones in the original box carry a price tag of fifty cents, c. 1925. $75-95.

Kum-a-part snap links with enamel in the original box, c. 1925. $75-95.

Art Deco tie bar and cufflink set in red and black enamel. The tie bar is marked "Swank" and the snap links "Kum-a-part". $150-200 set.

Art Deco snap links were offered in various color combinations. This design was available in red, blue, green, and white enamel, all in combination with black. $100-125 each.

Snap links with patterned designs, Kum-a-part, $50-75.

Snap links in silver metal with hammered and smooth finishes, Kum-a-part. $40-60.

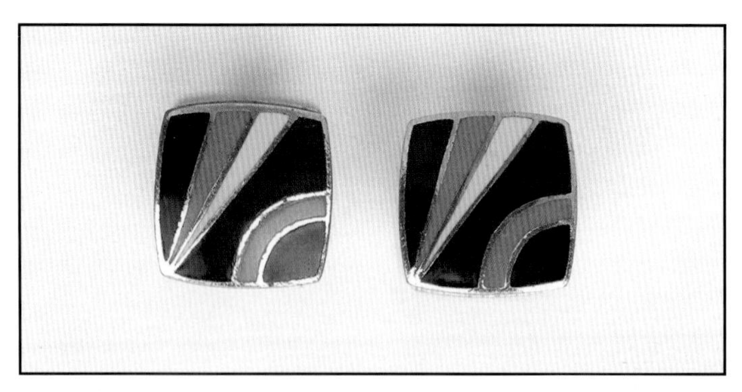

Art Deco enamel snap links, Kum-a-part, $100-150.

These enamel snap links show the guilloche process with a stamped pattern and translucent enamel. Kum-a-part by Baer & Wilde, $100-125.

Art Deco enamel snap links, Kum-a-part, $100-150.

Pearl and celluloid snappers, combined with patterned silver metal, Kum-a-part, $40-50.

Enamel snap links in black and white enamel, Kum-a-part, $75-100.

Enamel and pearl snap links, the latter set with blue stones to resemble Ceylon sapphires, Kum-a-part. $75-100.

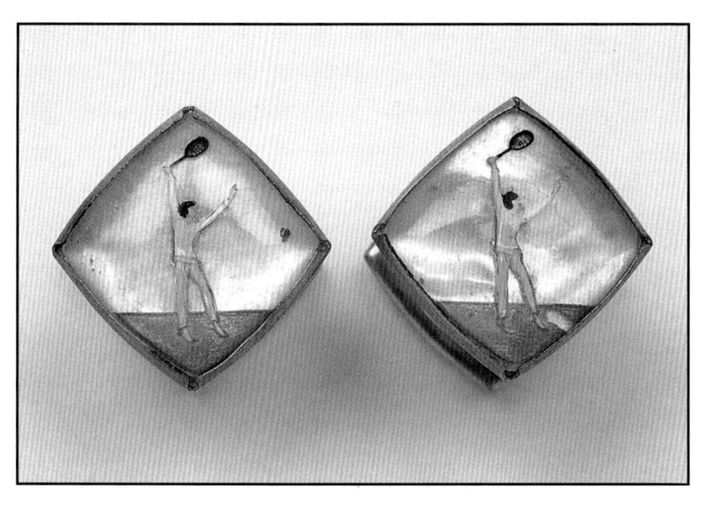

With scenes under glass, these tennis player snap links were designed after Essex crystal. $100-150.

Silver metal snap links show the influence of the Arts and Crafts period. $35-45.

Under glass snap links with horse head design showed the trend toward sporting jewelry, c. 1925. $100-150.

Stirrup cufflinks, Kum-a-part, silver and gold metal with stones. $50-65.

Mother of pearl snap links, Kum-a-part by Baer & Wilde, $25-50.

Snap links in enamel, mother of pearl, and celluloid, $35-65.

Sterling silver and celluloid Kum-a-part cufflinks, patented 1923, $60-75.

Kum-a-part cufflinks in traditional metals. $25-45.

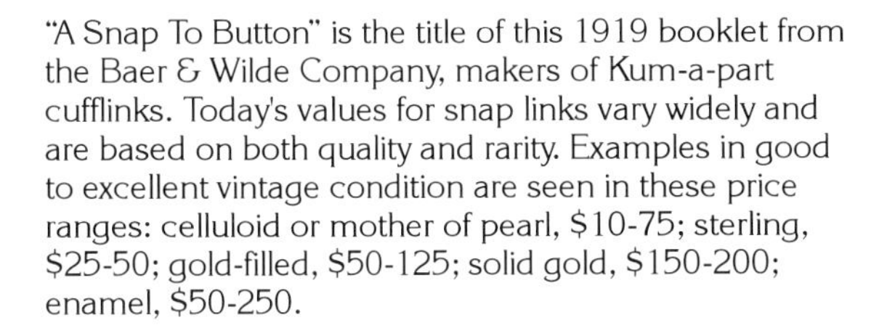

"A Snap To Button" is the title of this 1919 booklet from the Baer & Wilde Company, makers of Kum-a-part cufflinks. Today's values for snap links vary widely and are based on both quality and rarity. Examples in good to excellent vintage condition are seen in these price ranges: celluloid or mother of pearl, $10-75; sterling, $25-50; gold-filled, $50-125; solid gold, $150-200; enamel, $50-250.

An excerpt from the 1919 Baer and Wilde booklet, "A Snap to Button," says:

"'Elegance wedded to convenience.' Add to this modern utility all that artistry and skill can produce in distinction of pattern–and the countrywide appeal of Kum-a-part is explained.

"It is so up-to-date! So thoroughly in keeping with the spirit of the times. A man cannot help liking both the genius of its workmanship and its good looks.

"It is pleasing as a gift, delightful as a possession."

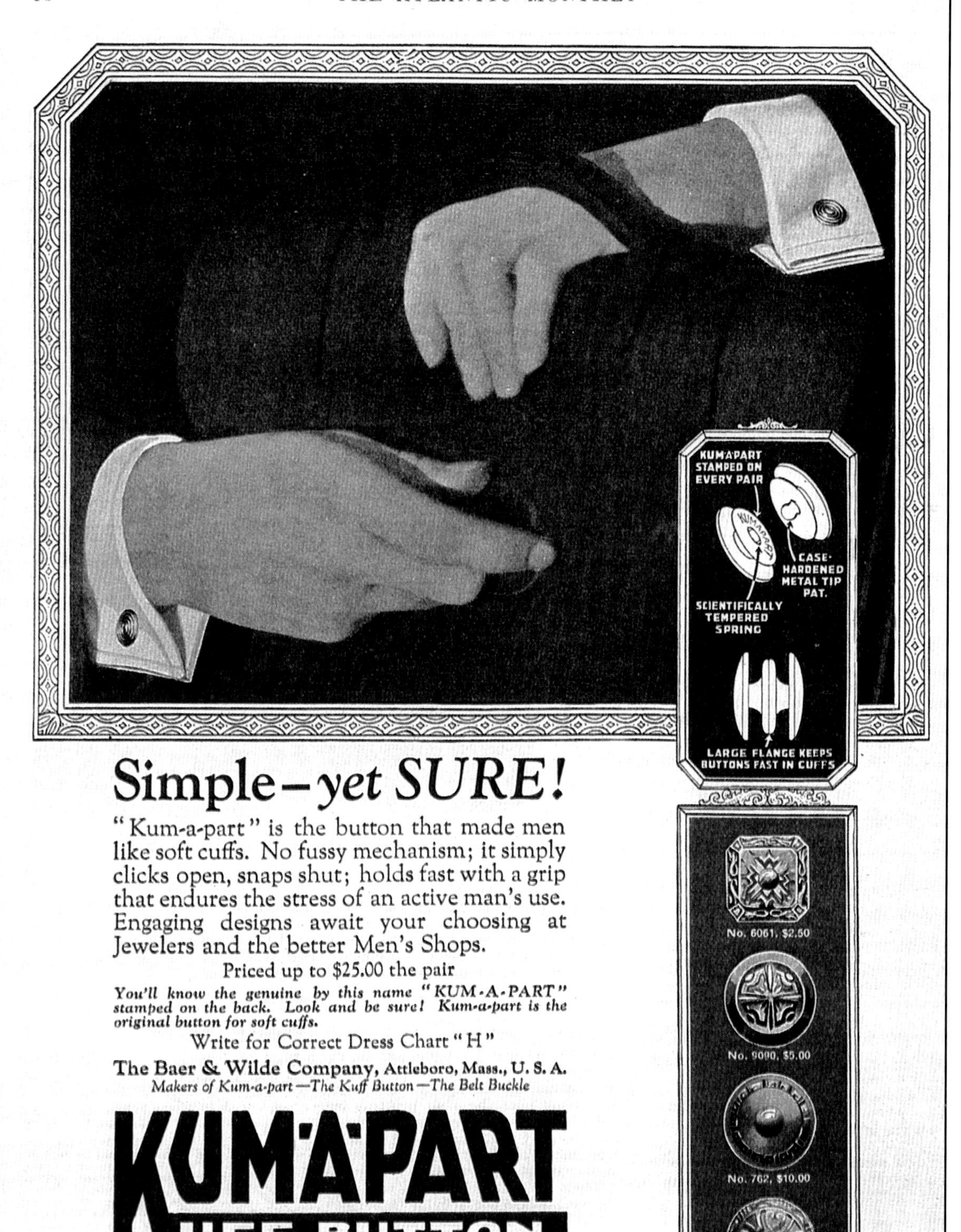

Magazine advertising for Baer & Wilde Kum-A-Part cufflinks.

Magazine advertising for Baer & Wilde Kum-A-Part cufflinks.

Magazine advertising for Baer & Wilde Kum-A-Part cufflinks.

StaLokt cufflink 1923 advertisement, J. F. Sturdy's Sons Company, illustrating the locking feature of this product

Chapter 9

Mesh Purses

1933-1934 Whiting and Davis Century of Progress, Chicago mesh purse, 3.5" x 4.5", plain silver clasped frame with rigid handle. This particular purse is green on white background. Century of Progress comet logo is noted. Mesh tiles large in size and indicative of the era. Many other primary color combinations exist. $600-750

1933-1934 Whiting and Davis Century of Progress Exposition, Chicago, Fort Dearborn mesh purse. Silver frame stamped on the front and back with "Century of Progress, Fort Dearborn 1833 Chicago 1933". This purse, 4" x 5.5", was exhibited in the General Exhibits Building, Aluminum Palace, Section 4 Exhibit 5 at the Worlds Fair in Chicago with Whitings' metal mesh making machines. $750-950

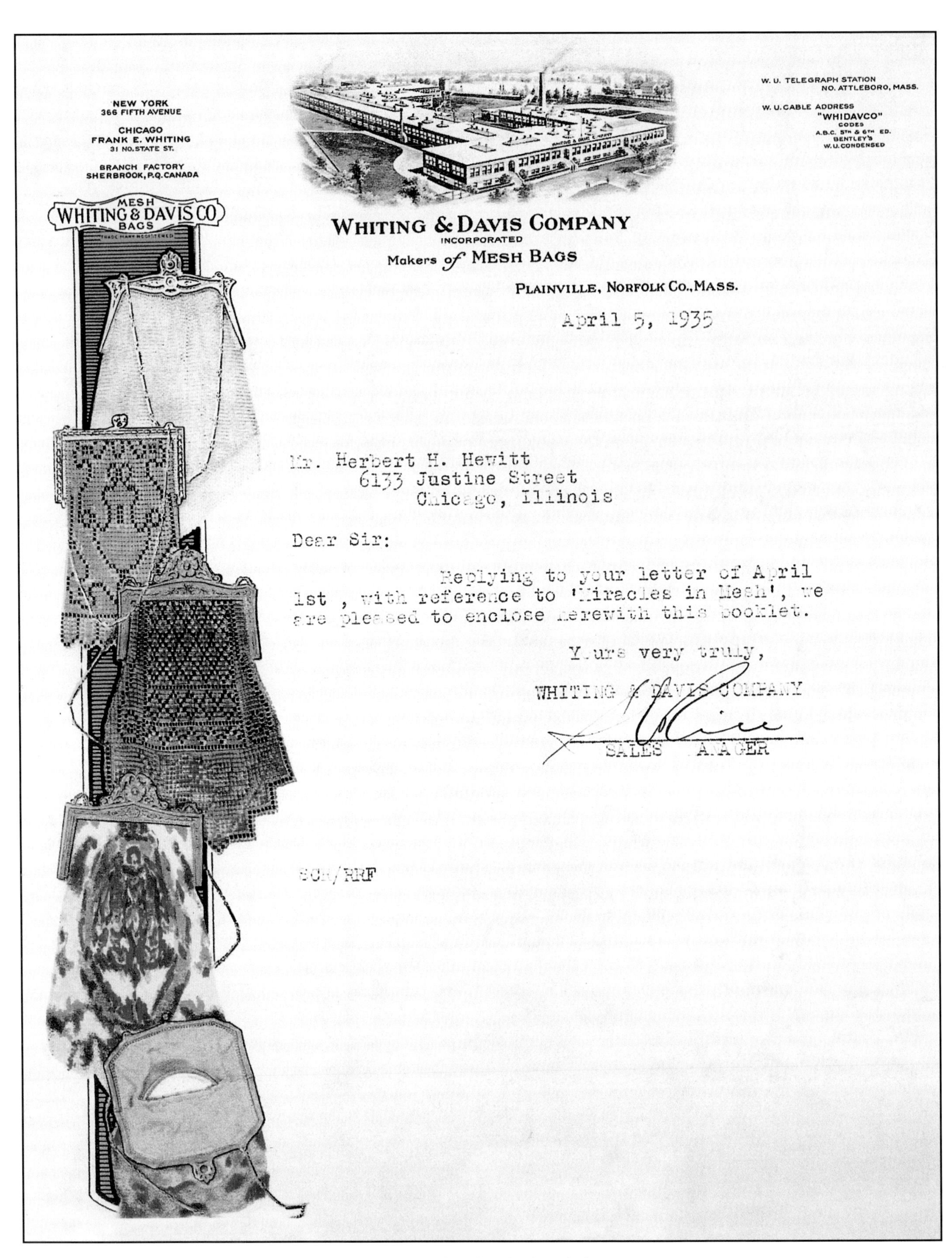

NEW YORK
366 FIFTH AVENUE
CHICAGO
FRANK E. WHITING
31 NO. STATE ST.
BRANCH FACTORY
SHERBROOK, P.Q. CANADA

W. U. TELEGRAPH STATION
NO. ATTLEBORO, MASS.
W. U. CABLE ADDRESS
"WHIDAVCO"
CODES
A.B.C. 5TH & 6TH ED.
BENTLEY'S
W. U. CONDENSED

MESH
WHITING & DAVIS CO
BAGS

WHITING & DAVIS COMPANY
INCORPORATED
Makers of MESH BAGS
PLAINVILLE, NORFOLK CO., MASS.

April 5, 1935

Mr. Herbert H. Hewitt
6133 Justine Street
Chicago, Illinois

Dear Sir:

Replying to your letter of April 1st, with reference to 'Miracles in Mesh', we are pleased to enclose herewith this booklet.

Yours very truly,

WHITING & DAVIS COMPANY

SALES MANAGER

SCH/RRF

Whiting and Davis letterhead acknowledging a request for the "Miracles in Mesh" brochure.

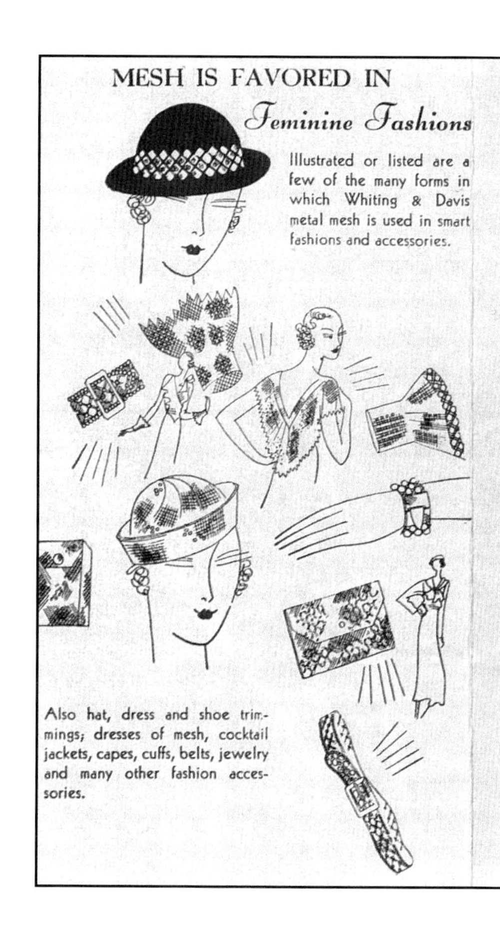

MESH IS FAVORED IN *Feminine Fashions*

Illustrated or listed are a few of the many forms in which Whiting & Davis metal mesh is used in smart fashions and accessories.

Also hat, dress and shoe trimmings; dresses of mesh, cocktail jackets, capes, cuffs, belts, jewelry and many other fashion accessories.

Miracles in Mesh in INDUSTRY AND SCIENCE

Metal mesh as produced by the Whiting & Davis Company has a wide range of uses in Industry and Science. The strength of metal woven into a fabric of almost silken texture, permits its use wherever endurance and flexibility are required.

- Diathermy Electrodes
- Motion Picture Screens
- Theatrical Effects
- Fireplace Screens
- Hotplate Pads
- Mudguard Flaps
- Blowout Patches
- Mesh Portieres
- Radiator Covers
- Lamp Shade Trimmings
- Drinking Glasses
- Picture Frames
- and many other articles.

MADE IN (WHITING & DAVIS CO. MESH BAGS) THE U. S. A

It is highly probable that there exist in your business ways in which metal mesh can be profitably used. For interesting information, write the

WHITING & DAVIS COMPANY
Plainville, (Norfolk County), Mass.

Printed in U.S.A.

CENTURY OF PROGRESS EXPOSITION ~ *Chicago 1933*

Manufactured and Exhibited by the
WHITING & DAVIS COMPANY
Aluminum Palace - Section 4 Exhibit 5
General Exhibits Building

THE *ROMANCE* OF *MESH* . . .

PLUMED knights jousting for a fair lady's favor—adventuring afar in quest of the Holy Grail—or loyally battling in defence of home and sovereign! Those are the romantic pictures which flash across one's mind at the mention of mesh, the ring-woven metal fabric from which ancient armor was made.

The exact period in which chain, or ring mesh armor was first used is unknown, but research reveals that it was at least prior to the year 700 A. D.

At first, only shirts or "hauberks," were made in ring mesh. Later, entire suits of armor were produced. The great labor and length of time required, made them very valuable, and by an ancient edict, they were "not to be bartered or sent over the seas."

There were several types of ring mesh, the armorers of different countries having varying ideas regarding the details of its making. Samples of English mesh armor made about the year 1000, for example, are heavier than mesh produced in France and other European countries. English mesh is also more intricate in its construction and should have had greater protective value than the French and other simpler types. Ancient methods of making metal mesh were very laborious. Iron wire was used, as steel was unknown. Wire of the required thickness was first wound about a round iron rod. When the rod was covered by these windings, each twist was cut with a chisel, thus making the individual "rings." Each iron ring was then flattened and shaped with a hammer so that the ends overlapped. Ring after ring was interlocked to form the desired garment, and as each tiny ring was added it was secured in place by *riveting its ends*. Thousands of rings! Many weeks of labor!

The advent of firearms ended the use of metal mesh as armor. The laborious hand method of manufacture of mesh for other uses, however, remained practically the same for nearly 200 years.

About the year 1909, the rapidly increasing demand for metal mesh for many purposes resulted in the invention of machines for its manufacture. These, in their latest developments, may be seen in the Whiting & Davis Company exhibit at the Century of Progress Exposition, Chicago, magically converting a bar of cold metal into shimmering, silken-textured metal mesh. The romance of mesh has become the modern miracle of mesh!

Miracles in Mesh . . .

IT is a long step from the cumbersome chain armor of olden days to the delicacy and charm of the many articles now produced in metal mesh—colorful creations of rare beauty, deftly fashioned into fascinating feminine allurements—truly Miracles in Mesh!

In Industry and Science also, metal mesh plays important exclusive parts, practical, experimental,—far removed from the day and need of its origin.

To one man in America is credit principally due for these Miracles in Mesh. Mr. Charles A. Whiting, President of the Whiting & Davis Company, long ago foresaw its possibilities in Fashion, Industry, Science.

The marvelous machines shown in the Whiting & Davis Company exhibit at the Chicago Century of Progress Exposition are the creations of his vision and determination—machines which produce in a day more metal mesh than the ancient armorer could weld in a lifetime.

On the following pages are shown or listed many modern uses of these Miracles in Mesh!

Whiting & Davis Company, Plainville (Norfolk County), Mass.

The "Miracles in Mesh" brochure from the Century of Progress Exposition in Chicago, 1933-1934, offers a history of Whiting and Davis and mesh.

ChapterWhiting and Davis, large flat tile mesh purse with desirable scenic silhouette of romantic Victorian couple. Interesting complimentary enameled mesh frame. 6" x 8". Scarce, $600-800.

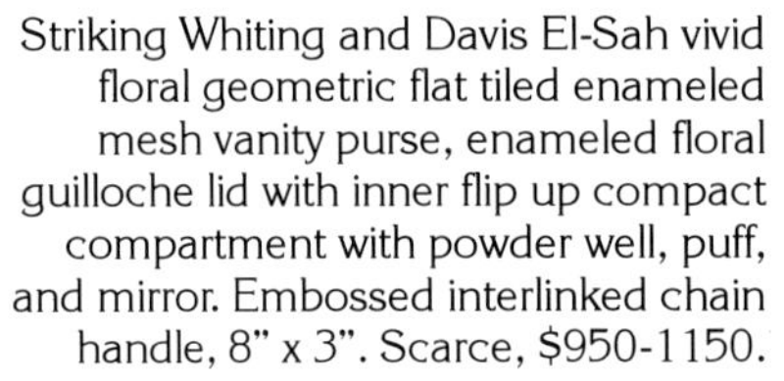

Striking Whiting and Davis El-Sah vivid floral geometric flat tiled enameled mesh vanity purse, enameled floral guilloche lid with inner flip up compact compartment with powder well, puff, and mirror. Embossed interlinked chain handle, 8" x 3". Scarce, $950-1150.

IRENE RICH, talented star of many stage and screen triumphs knows—like all women with a sense of the exquisite—that after the last finishing touch, she goes forth more confident if a smart and lovely Whiting & Davis Mesh Bag graces her costume. Her most *useful* article of dress, too. Mesh bag illustrated is the Filigree design.

WHITING & DAVIS CO.
Plainville (Norfolk County), Mass.
In Canada, Sherbrooke, Que.

Whiting & Davis Mesh Bags

In the Better Grades. Made of the Famous "Whiting" Soldered Mesh

Advertisement from a 1920s *Photoplay* magazine shows stage and screen actress Irene Rich.

Movie stars traveled to Plainville, Massachusetts to receive complimentary mesh purses from the company's president, Charles A. Whiting. Their fancy automobiles lined up at Whiting and Davis, while the actresses enjoyed their visit.

Compact purse by Charles Thomae Co. with guilloche enamel hand painted top, tiny and fine enameled mesh, beautifully detailed frame, cabochon sapphire in the clasp, and woven silver chain handle, marked with "crown, T, and lion" and "3442-Sterling". Scarce, $2000-3000.

Mickey Mouse child's purse by Whiting & Davis, $300-400.

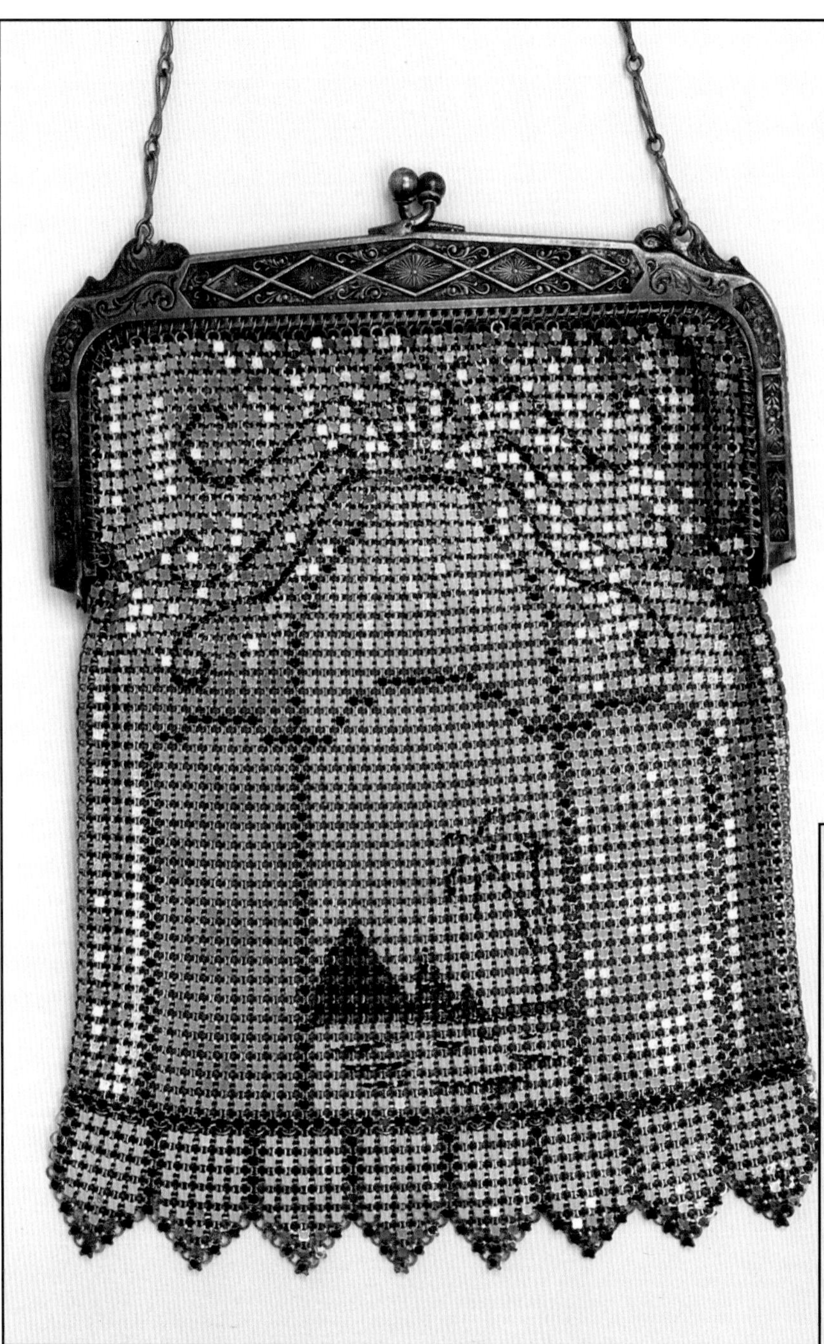

Egyptian motif purse, Whiting & Davis, c. 1925. $300-400.

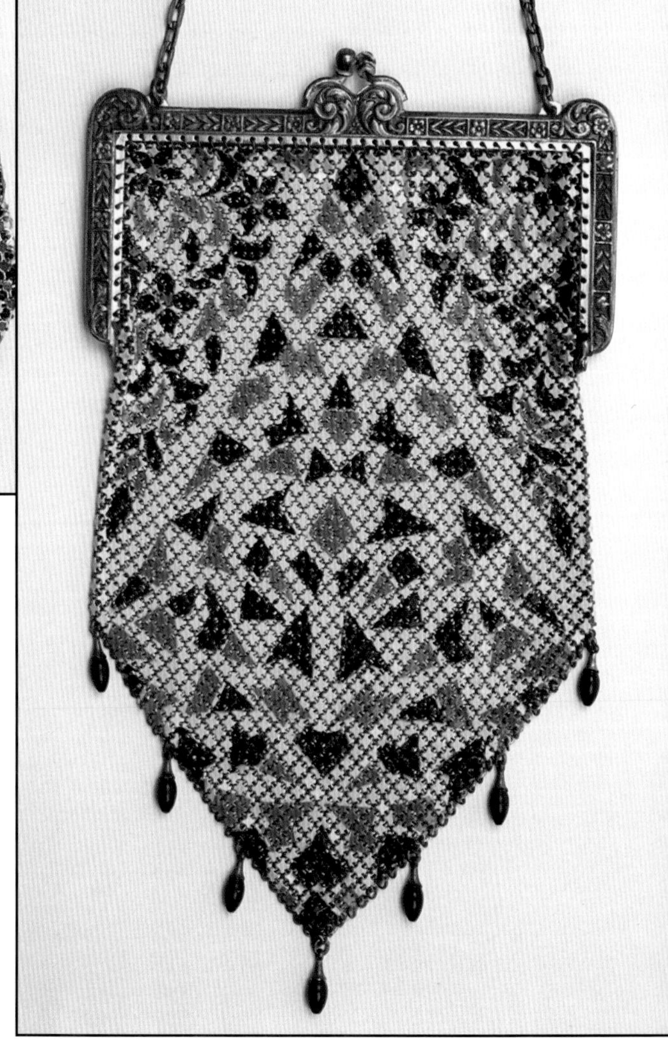

Art Deco enamel purse by Mandalian, orange and black, accented by oval black enamel drops, c. 1925. $300-450.

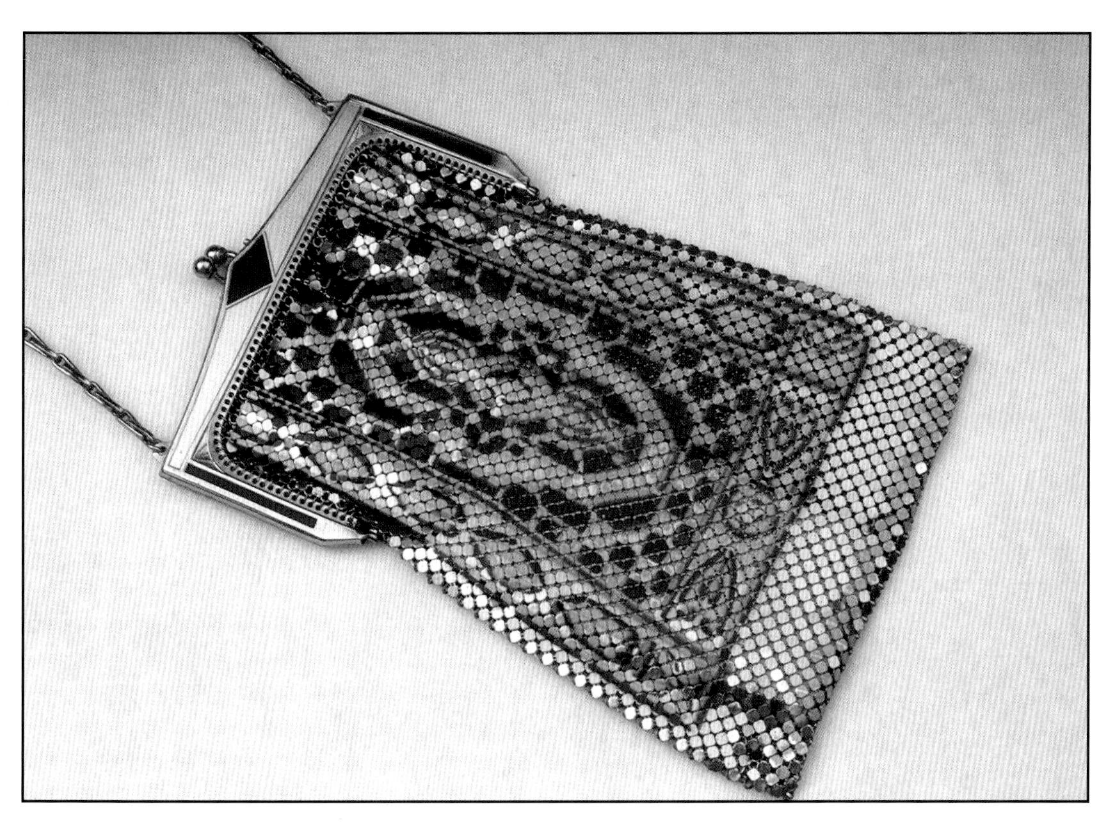

Art Deco enamel purse, Whiting & Davis, c. 1930. $300-400.

Egyptian Revival purse, Whiting & Davis, c.1925. $400-500.

Peacock purse by Mandalian, c. 1925. $500-600.

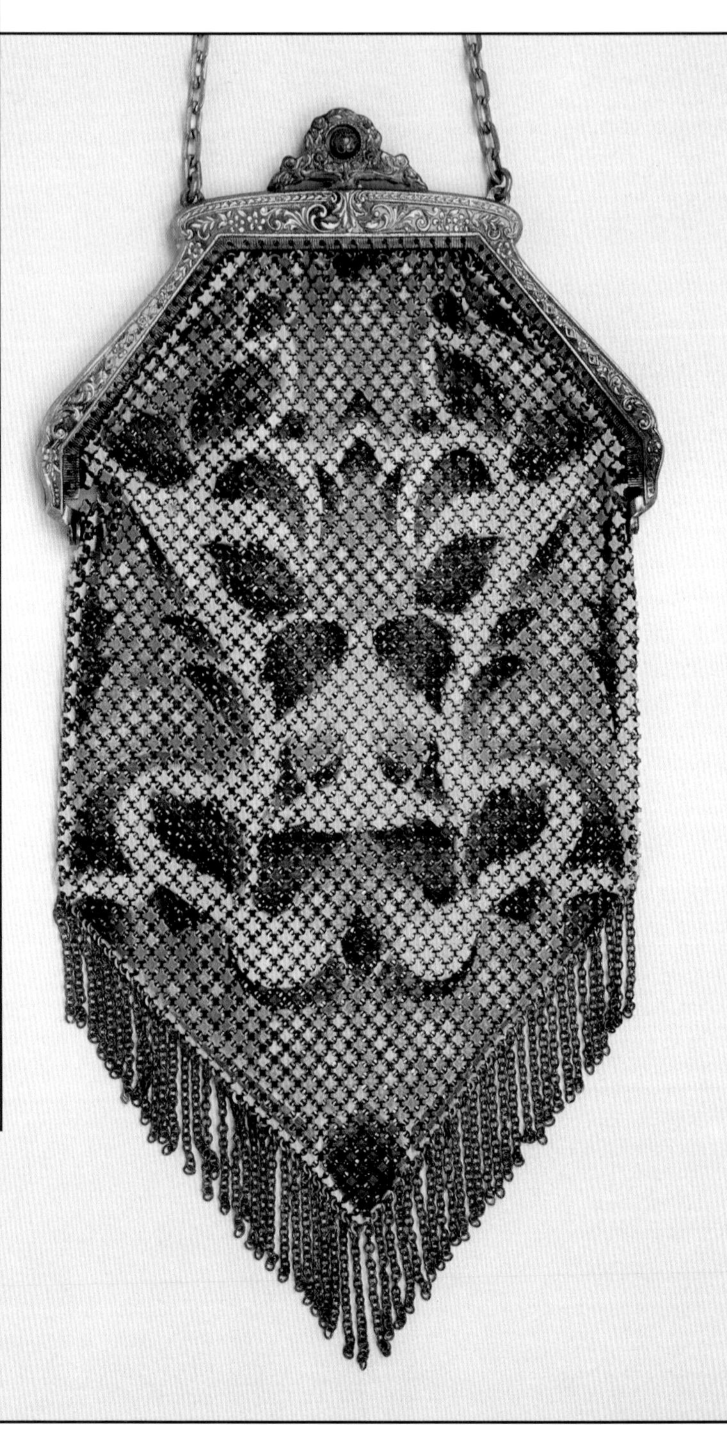

Teal blue-green purse by Mandalian, c. 1925 $300-450.

Dresden Mesh purse, Whiting and Davis. $350-500.

Dresden Mesh purse, Whiting and Davis. $450-550.

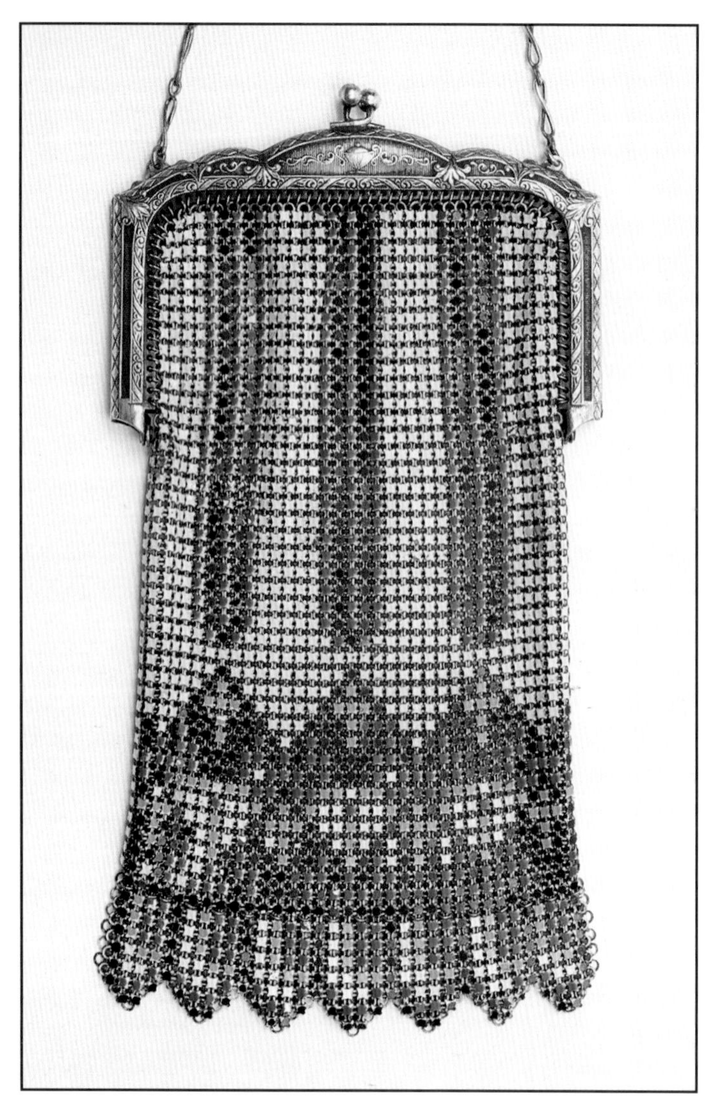

Art Deco enamel mesh purse, Whiting and Davis $300-400.

Roses purse, Mandalian $375-475.

The soft colors of turquoise and pink were prevalent in the period, as seen on this Mandalian purse of the early 1920s. $500-750.

Art Deco enamel mesh purse in "new condition" with original Whiting & Davis tag. $450-550.

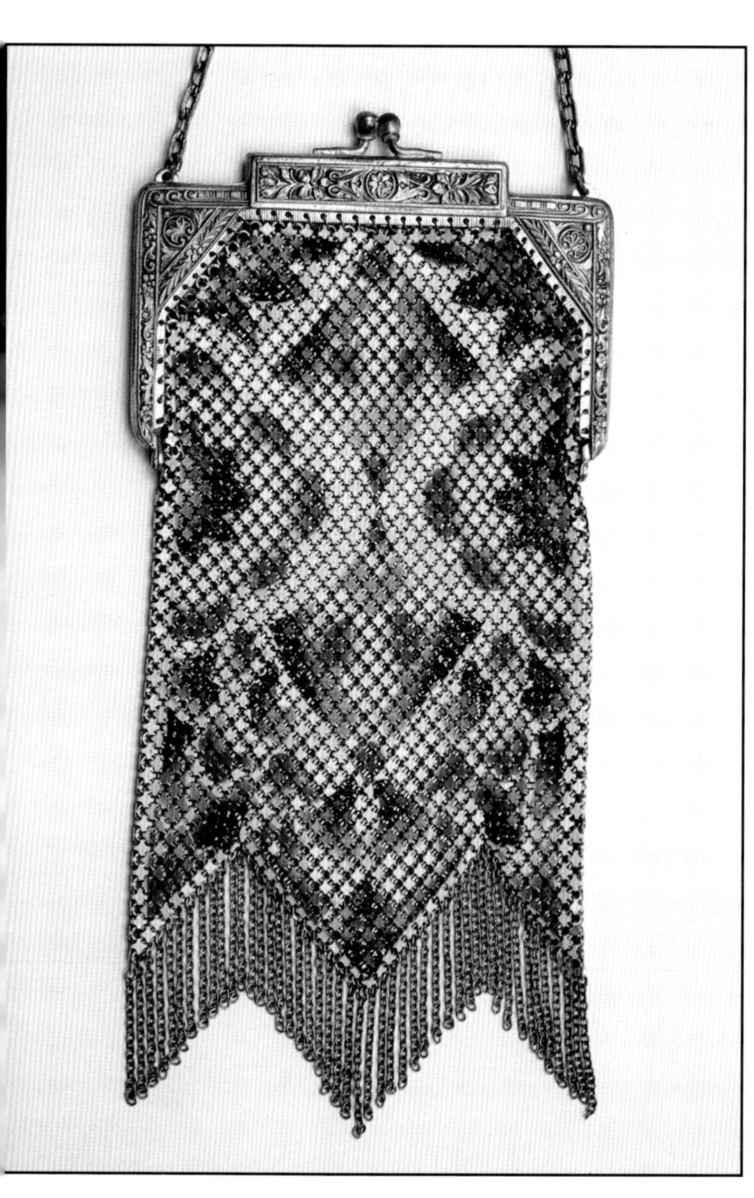

Art Deco Mandalian purse, c. 1925. $350-450.

Rose purse by Whiting and Davis, c. 1925, exceptional gold metal filigree frame. $350-500.

Whiting and Davis 1927 Enamel Mesh Purse Advertisements.

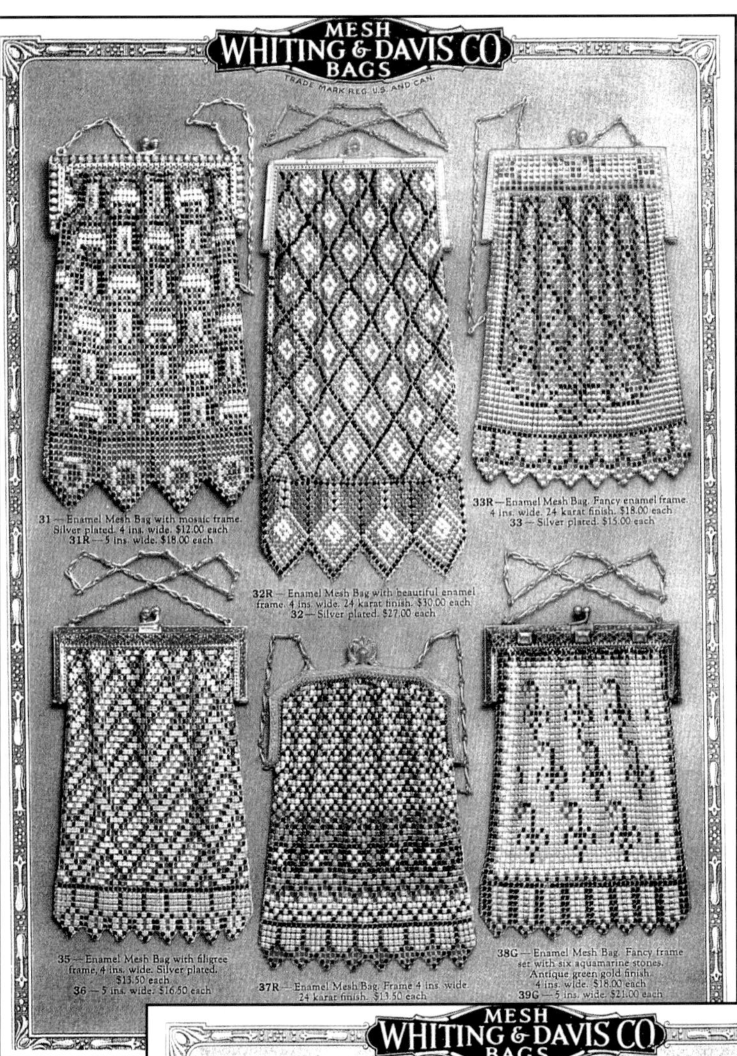

Whiting and Davis 1927 Enamel Mesh Purse Advertisements.

Hand in Hand *with Fashion*

"For Gifts That Last
WHITING & DAVIS CO
Consult Your Jeweler"

COLOR photograph illustrates enameled Costume Bag, silk lined, with pocket and mirror. Antique gold-finished frame, five inches wide

FREE
PORTFOLIO

ASK your jeweler to show you our very newest bags of MODERNIST DESIGN. Write us for *free portfolio* showing twenty-four costume bags in color

NATIONAL BROADCAST . . . simultaneous in Park Avenue and Pasadena . . . in New York and your town . . . the new Whiting & Davis Enameled Costume Bags. They go hand in hand with Fashion . . . hand in hand with those who know just how to give a flick of personal distinction and originality to the basic fashion for gay costume colors.

Always Whiting & Davis Enameled Costume Bags have that *precious* quality which only jeweler-craftsmanship bestows . . . and wherever fine shops are, you may buy them now. Like a broadcast of color harmonies they have gone out to jewelers and jewelry departments from coast to coast. Write for 1928 portfolio . . . order the pattern you prefer from your local jeweler.

WHITING & DAVIS COMPANY *World's Largest Manufacturers of COSTUME BAGS*
WITH WHICH IS ASSOCIATED WHITING & DAVIS CHAIN CO. MAKERS OF COSTUME JEWELRY FOR EVERYONE
PLAINVILLE (NORFOLK COUNTY) MASSACHUSETTS
In Canada: SHERBROOKE, QUEBEC

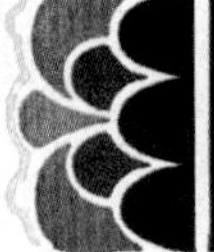

Whiting & Davis Costume Bags

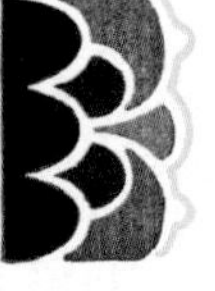

Whiting & Davis "Hand In Hand With Fashion" Advertisement of 1928.

Whiting & Davis Enamel Mesh
Purses "Portfolio of 1928." .

Whiting & Davis Enamel Mesh Purses "Portfolio of 1928." .

Whiting & Davis Enamel Mesh
Purses "Portfolio of 1928." .

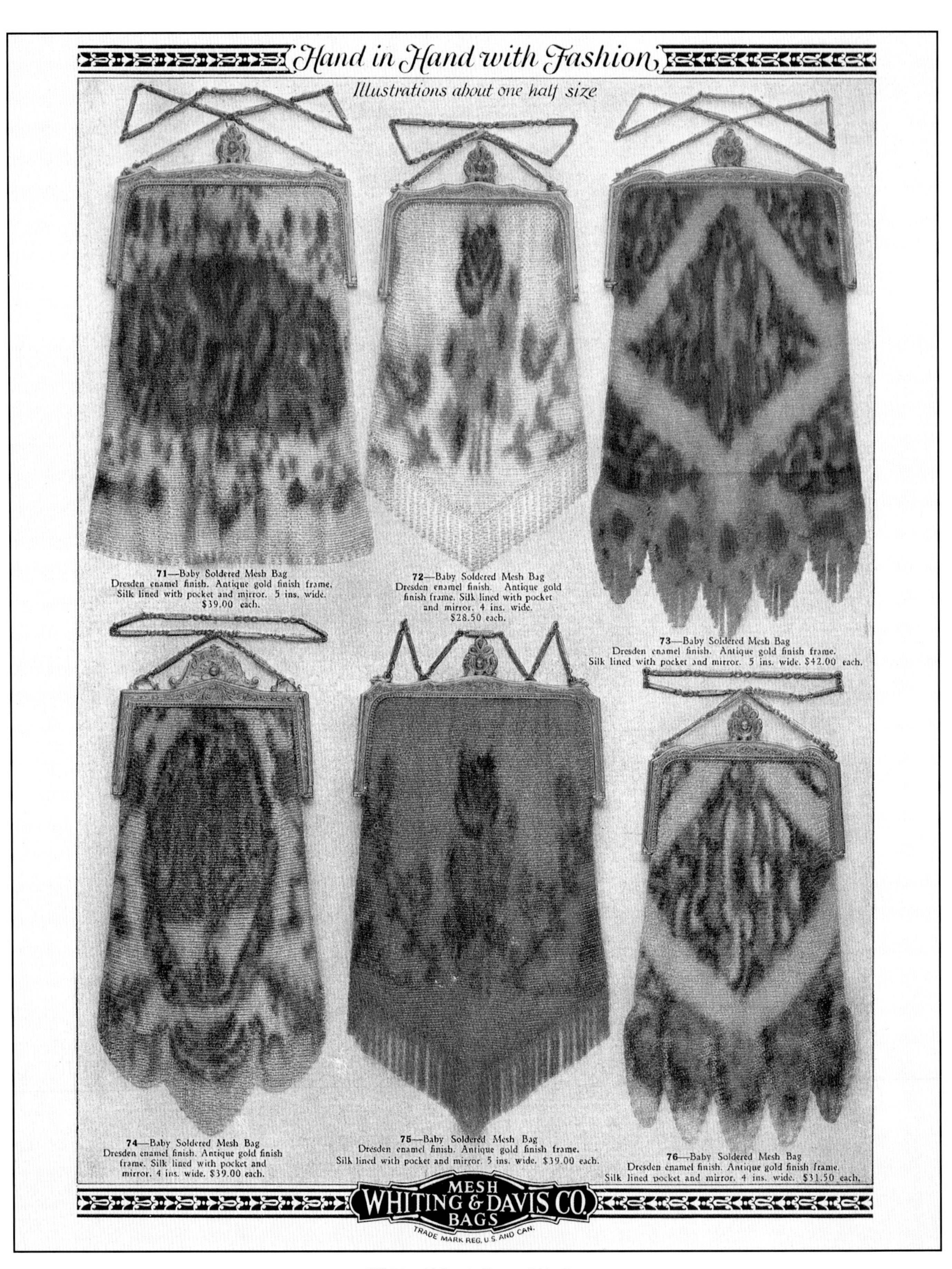

Whiting & Davis Enamel Mesh
Purses "Portfolio of 1928." .

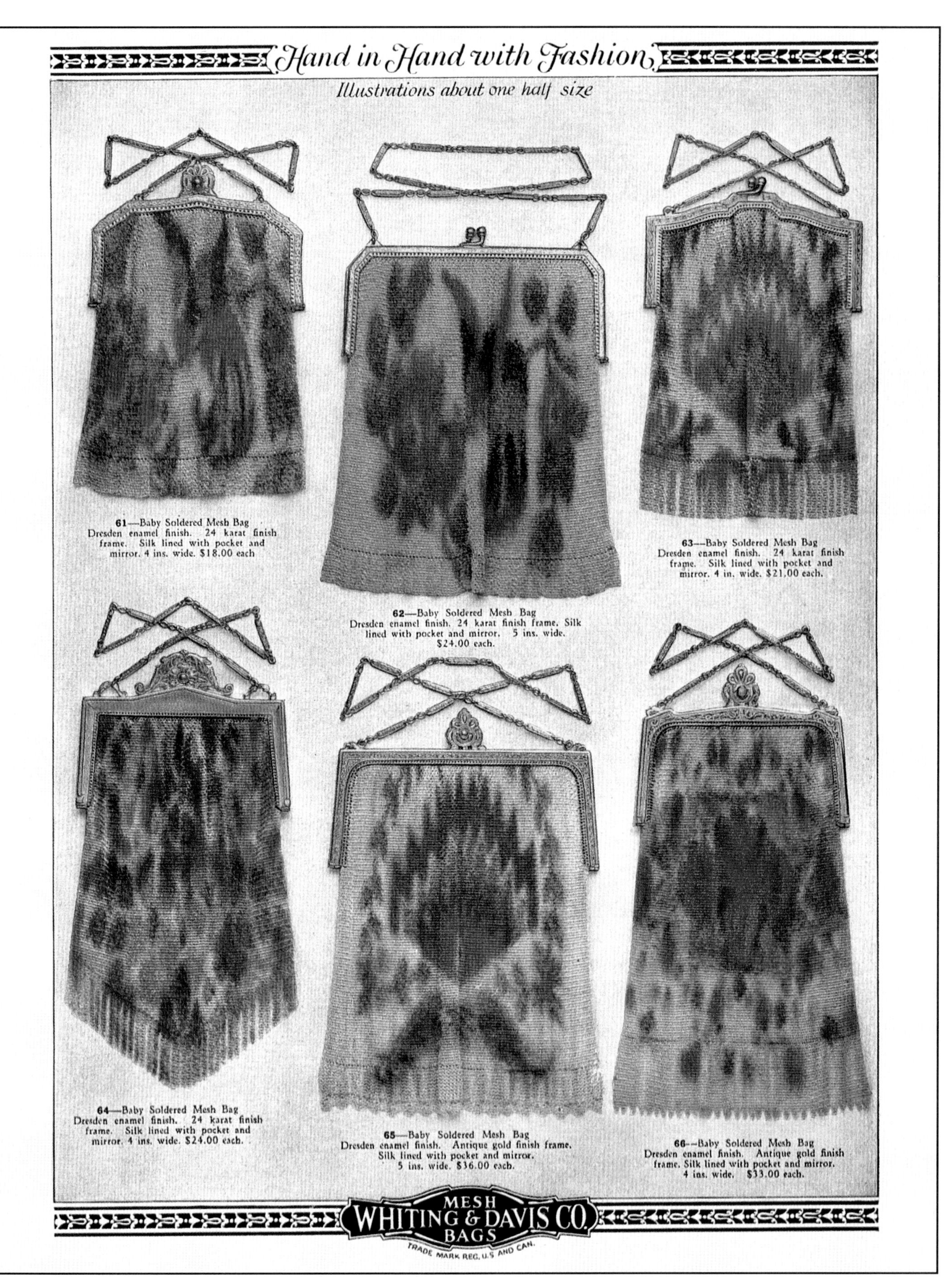

Whiting & Davis Enamel Mesh Purses "Portfolio of 1928." .

Whiting & Davis 1931 Mesh Purse Advertisement.

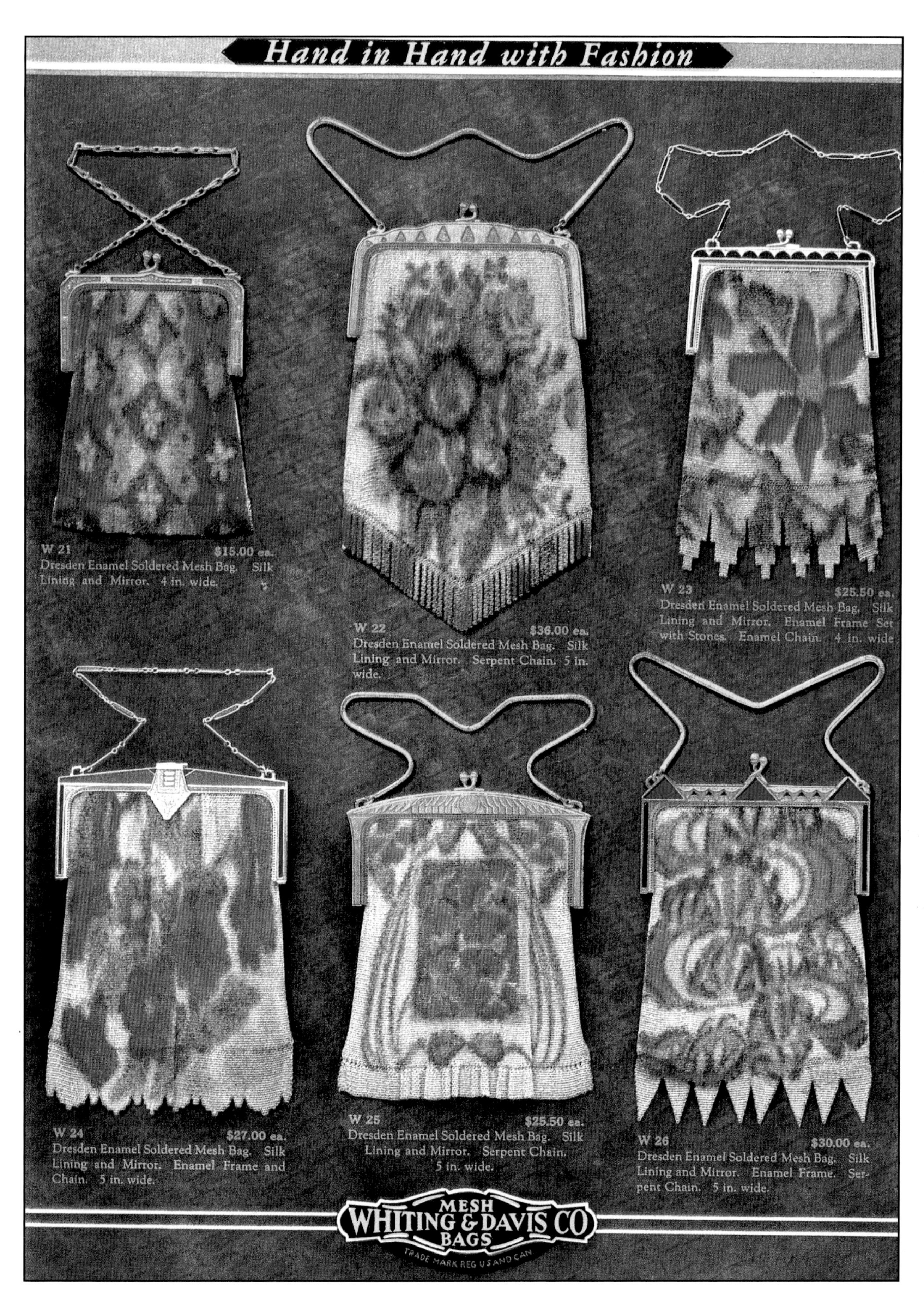

Whiting & Davis 1931 Mesh Purse Advertisement.

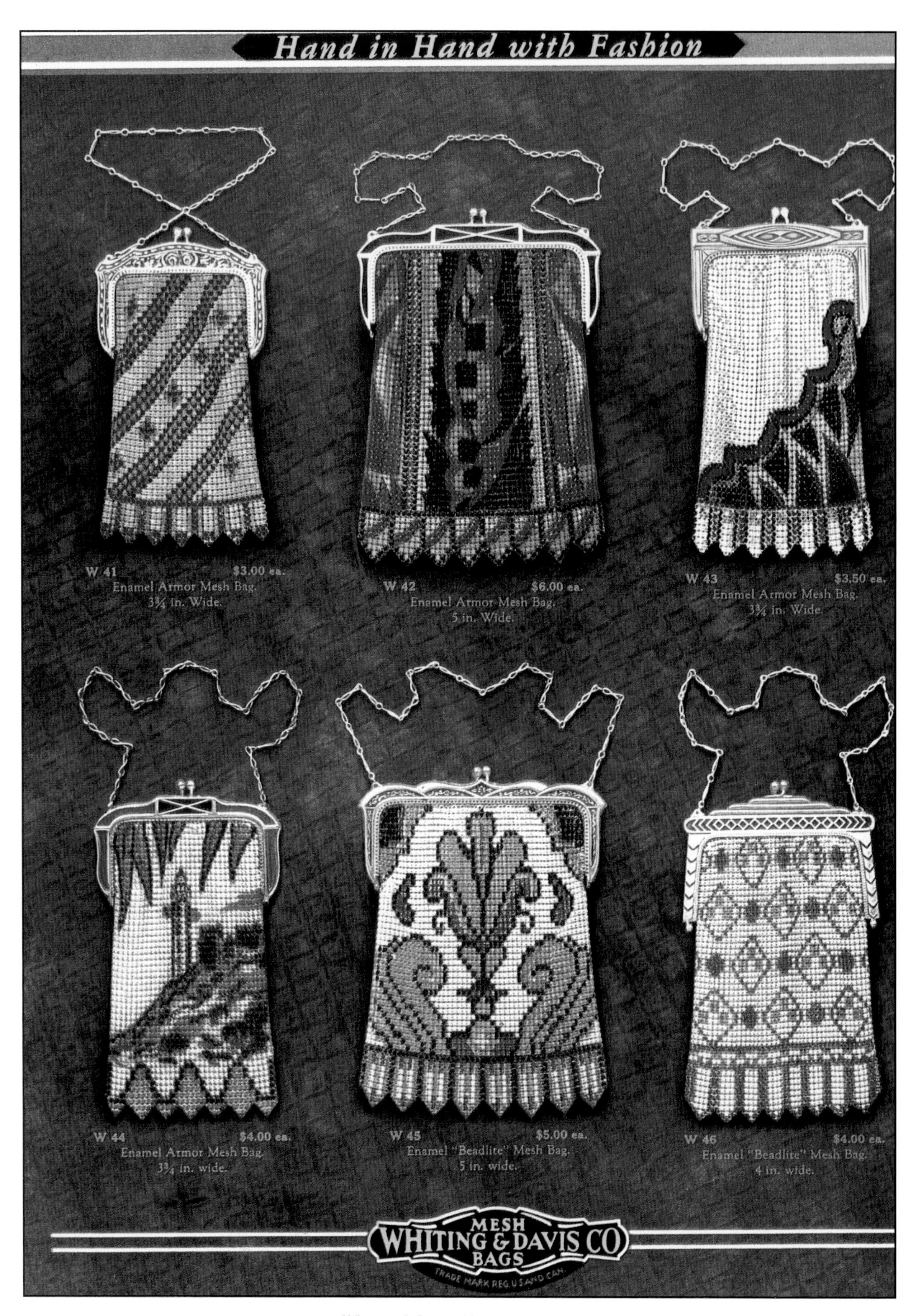

Whiting & Davis 1931 Mesh Purse Advertisement.

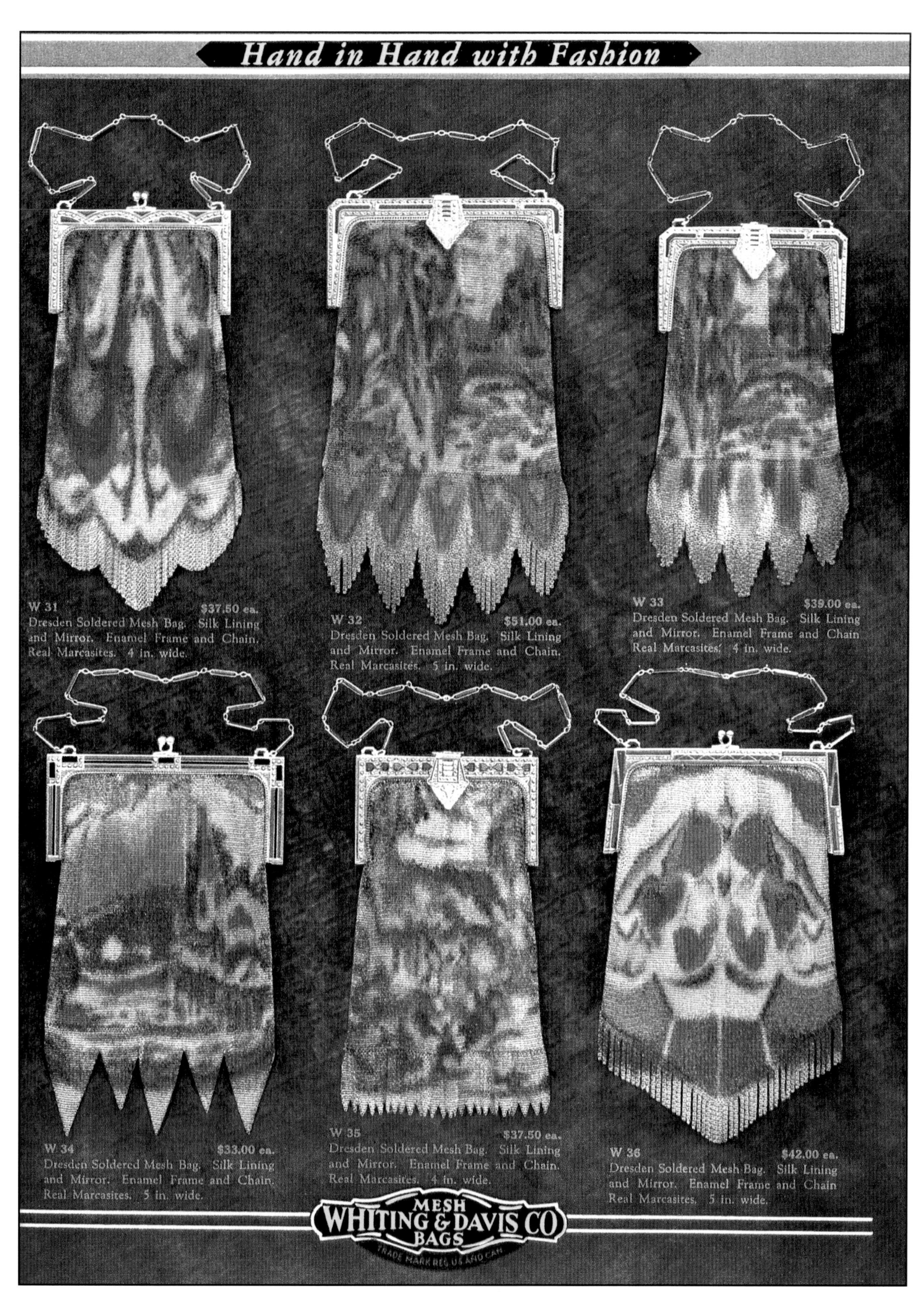

Whiting & Davis 1931 Mesh Purse Advertisement.

Chapter 10

Evans is Elegance

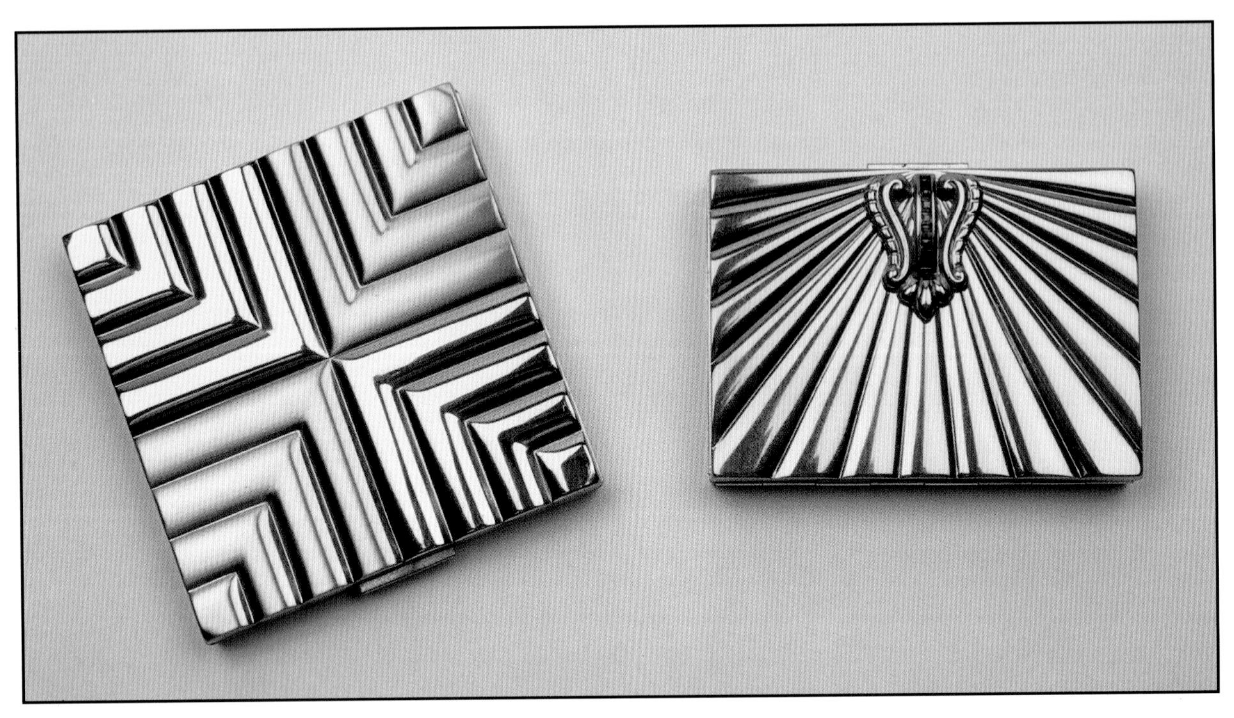

Compacts by Evans Case Company of North Attleboro, Massachusetts—whose trademark, "Evans is Elegance," indicated the quality of their products; gold tone metal, $50-75 each.

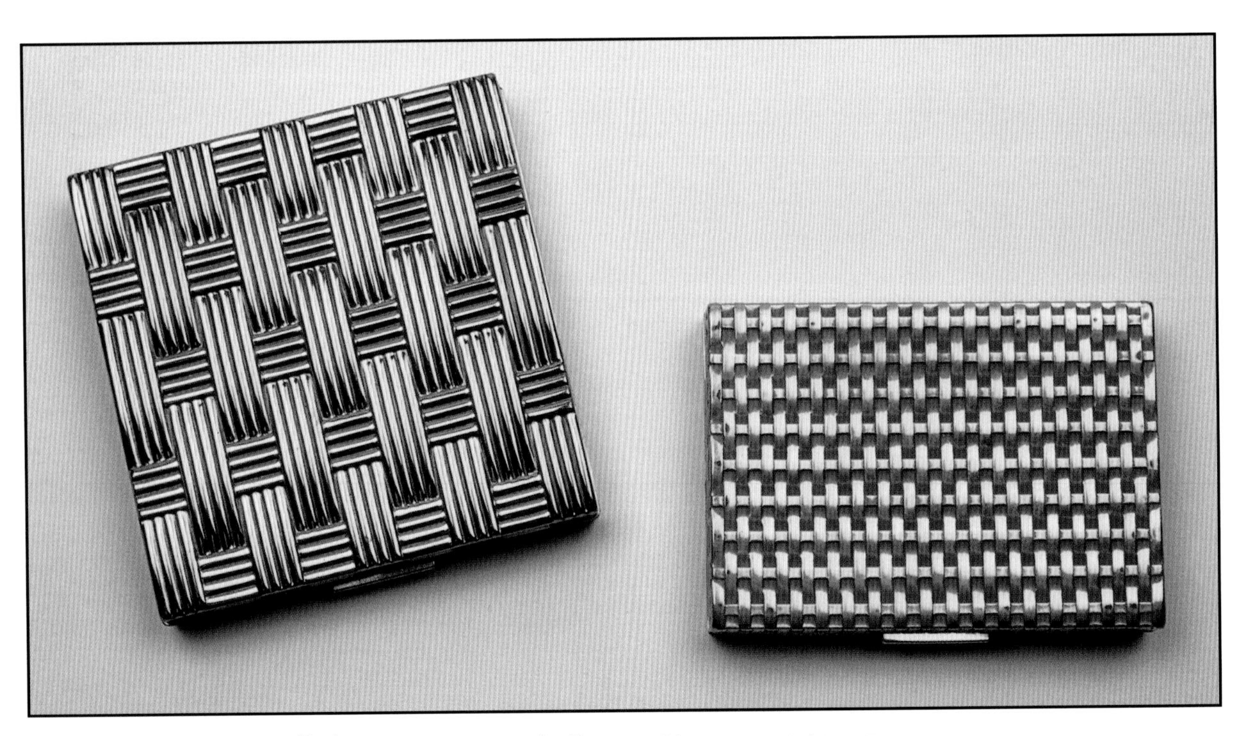

Basket weave compacts by Evans, gold tone metal, $40-65 each.

Compacts with enamel and rhinestones by Evans, $50-100.

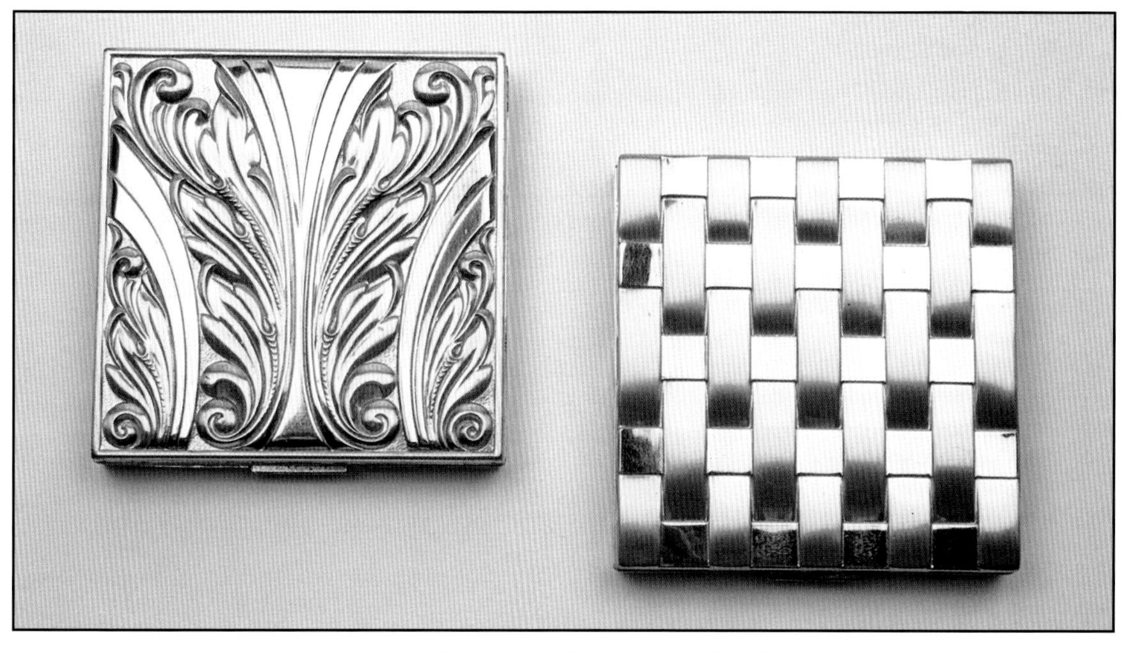

Compacts by Evans, gold tone metal. $40-65.

Compact with timepiece by Evans, $85-125.

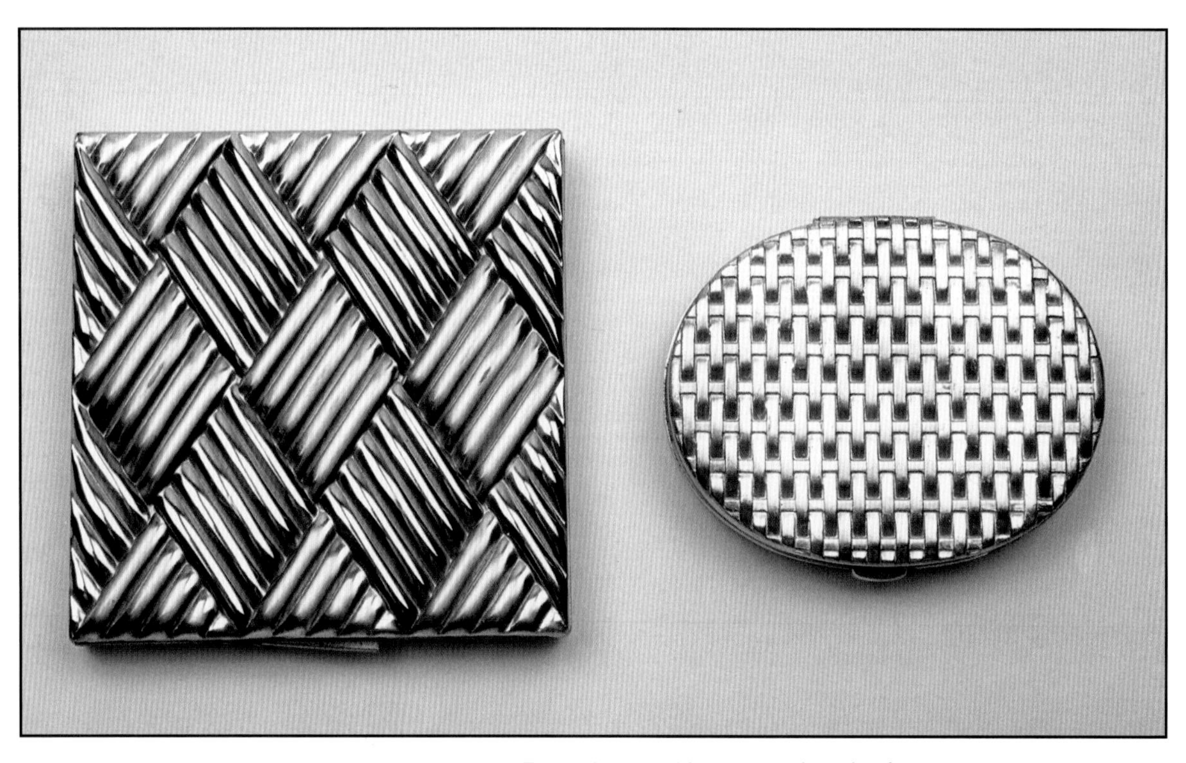

Basket weave compacts by Evans; left is gold tone metal, and right is sterling silver with rose and yellow gold. Left: $75-95; right: $100-150.

Cigarette case, gold tone metal with red
rhinestones to resemble rubies. $100-125.

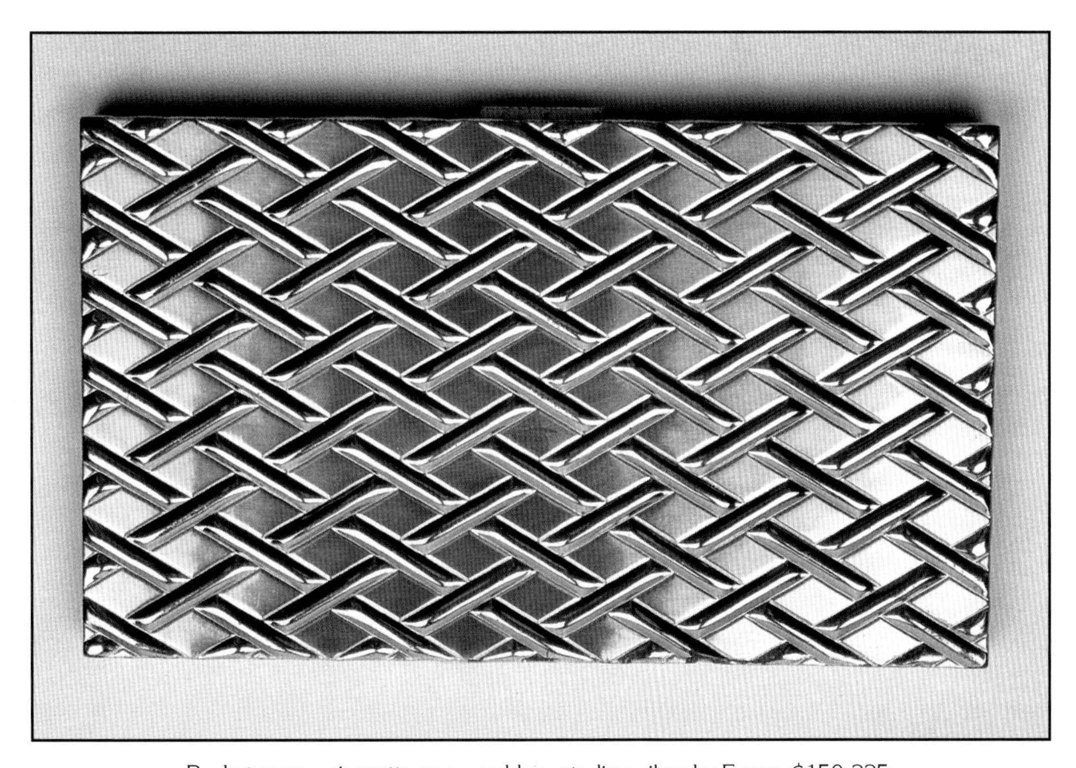

Basket weave cigarette case, gold on sterling silver by Evans. $150-225.

Heart compact in gold tone metal by Evans. $75-95.

Compact in gold on sterling silver by Evans. $95-125.

Lighters with stones and enamel by Evans. $35-55.

Sweetheart compact from World War II by Evans. $60-75.

Large gold on sterling compact by Evans, 4.75" diameter, $150-225.

Enamel compact by Evans, $95-125.

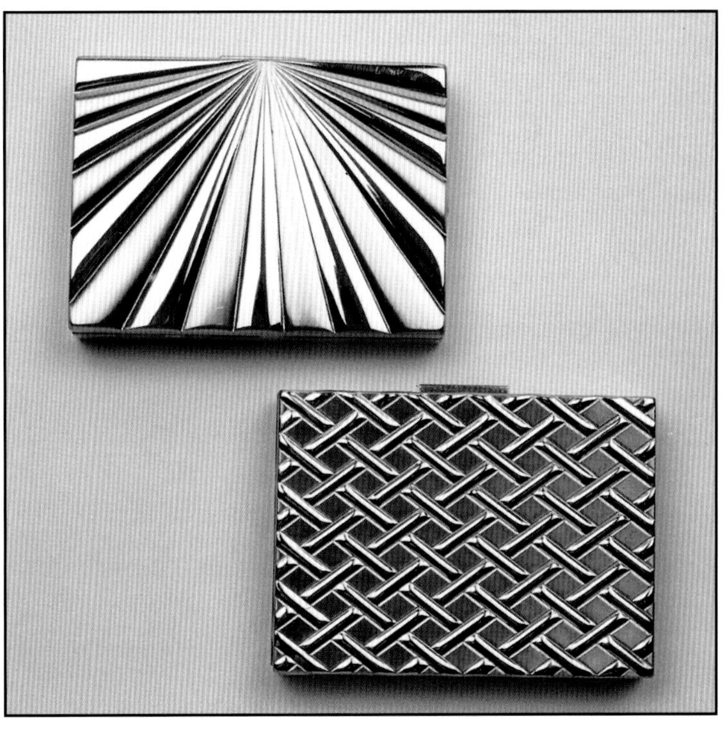

Compacts in gold tone metal by Evans. $45-75.

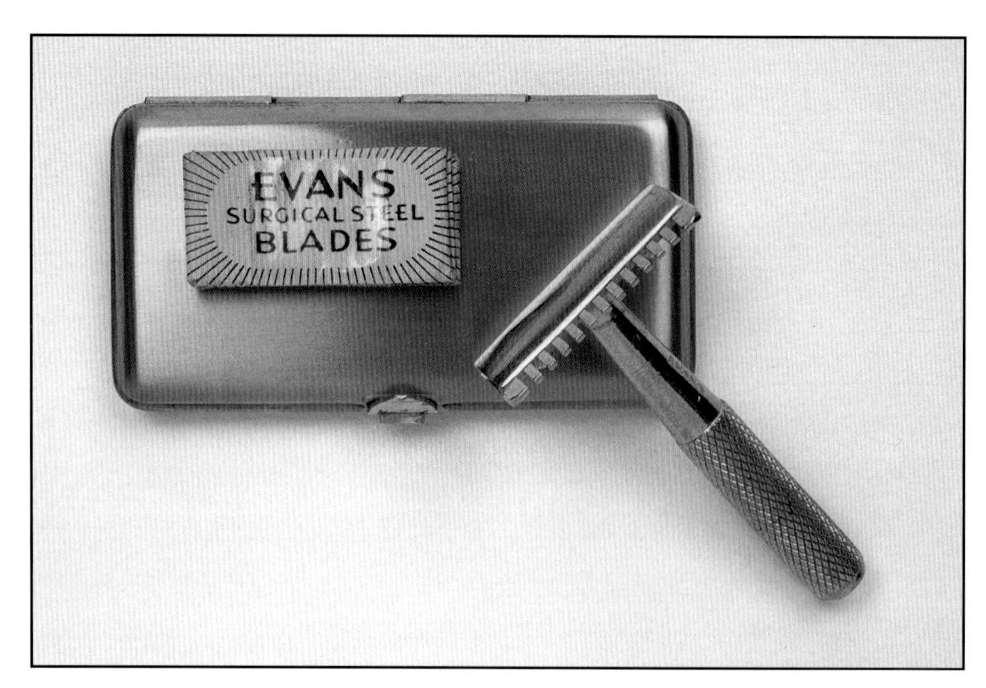

Miniature shaving set by Evans. $25-35.

Guilloche enamel brown and white necklace and earring set by Evans, $200-300.

Enamel snap cufflinks, attributed to Evans Case Company. It is also known that Swank offered a revival of snap cuff links, known today as "snappers", c. 1955 after the design of their world famous Kum-a-part cufflink. The style of enamel work on these cuff links is indicative of Evans, although they are unmarked. $50-75.

Guilloche enamel earrings in turquoise, Evans, $50-75.

Guilloche necklace and earring set in light blue enamel by Evans, $200-300.

Bracelets by Evans from the 1950s show modern influences. $60-85 each

Guilloche necklace and earring suite by Evans in light pink enamel from the 1950s. An elegant women of the time could also purchase an automobile in a similar color, equipped with accessories by the Evans Case Company. The Dodge La Femme, introduced in January of 1955, was a Custom Royal Lancer Hardtop, with an exterior color of Heather Rose over Sapphire White.A gold "La Femme" nameplate on each fender replaced the Royal Lancer nameplate. The interior was a Heather Rose Cordigrain bolster and trim. Designed for the fashionable woman, this automobile included a shoulder bag in soft rose leather fitted in a special compartment attached to the back of the passenger's seat—the purse included a compact, lipstick, comb, coin purse, cigarette case, and lighter in coordinating pink leather. The pink shoulder bag and its matching accessories were manufactured by the Evans Case Company of North Attleboro, Massachusetts. This necklace and earring set would have been an appropriate complement, c 1955. $250-350.

Chapter 11

Attleboro Jewelery Makers and their Marks

Some of the early names associated with jewelry manufacturing in Attleboro were Robinson, Draper, Richards, Bacon, Daggett, Codding, Tifft, Whiting, Sturdy, and Hayward. The nineteenth century was a time of tremendous growth for the jewelry industry in Attleboro. The following information about the earliest makers is culled from *History of Massachusetts Industries-Their Inception, Growth, and Success-Volume 1* by Orra L. Stone.

About 1807 Col. Obed Robinson began jewelry manufacturing in a part of Attleboro known as Robinsonville. David Brown, a highly skilled jeweler, was employed by Robinson and is believed to have learned the jewelry trade from "The Frenchman," for whom he worked in earlier years.

In 1810 Manning Richards built a small shop on his Cumberland Road farm and operated a successful jewelry manufacturing business there for many years.

In 1821 Draper, Tifft, and Company built a two-story factory, which originally measured 22 x 40 feet, being enlarged numerous times in later years. With many changes in partnership, the firm grew to become the largest jewelry manufacturer in the United States. In 1862 it became J. F. Bacon and Company.

In 1830 Calvin Richards and George Price built a jewelry factory. Known as Richards and Price, they later formed a partnership with S. D. Daggett.

In 1831 Dennis Everett and Otis Stanley opened a business for making watch chains and keys in South Attleboro.

In 1833 H. M. and E. I. Richards began jewelry manufacturing with numerous partners, including George Morse, Virgil Draper, and Abiel Codding. The firm was called Ira Richards and Company, and renamed E. Ira Richards in 1875.

In 1836 or 1837 Stephen Richardson and Company began making jewelry, joined by Abiel Codding. The firm was the first to export its products to Europe. It was also the first to open a permanent office in New York City, in 1854. In the 1870s, the company began trade with Japan, again setting a new precedent.

In 1837 Daggett and Robinson built a brick shop in West Attleboro, becoming Robinson and Company in 1850. This firm was carried on by members of the Robinson family throughout the century.

In 1840 Albert C. Tifft and William D. Whiting opened a jewelry business in the rear room of a blacksmith shop. Being successful from the beginning, they built a two-story brick factory the following year. In 1847, they gained the water privilege of the Beaver Dam and built a stone factory, which was enlarged numerous times to accommodate the production of silverware, as well as gold products. In 1853 Mr. Whiting purchased the interests of Mr. Tifft and renamed the firm W. D. Whiting and Company. This firm was the last of the large silver manufacturers to use the apprentice system, where a boy agreed to work for a four-year apprenticeship, earning 8 cents an hour to begin, and gradually increasing to 14 cents an hour. In 1876 the company moved to New York.

In 1847 J.J. Freeman and B. S. Freeman opened a business in their father's house, and in 1851 built a factory in Attleboro Falls. They specialized in gold jewelry production under the names Freeman and Brother, and later Freeman and Company. The firm was the first to manufacture curbed chains of rolled gold in the United States. J. J. Freeman purchased an imported gold chain and after extensive study invented a machine to manufacture curb chain, replacing hand labor.

In 1849 John F. and James H. Sturdy moved their jewelry manufacturing company from Providence to Attleboro. They devised a method of making rolled and stock plate and introduced this production to Attleboro, along with their partner Herbert M. Draper, in the firm of Draper, Sturdy, and Company. They were philanthropists in the way they shared production secrets for manufacturing rolled gold plate stock and jewelry with other Attleboro jewelry makers, and this method soon became universal in the area. The company was later renamed J. F. Sturdy's Sons Co.

F. G. Whitney and E. W. Davenport started the F. G. Whitney Company, which manufactured gold plated jewelry throughout the century, establishing a large foreign business. The firm became the Walter E. Hayward Company in 1891, famous for it's gold front, gold plated, and gold-filled jewelry, which was shipped to Canada, the Philippines, South America, China, and Japan.

Ackerman Bros.
(K 1931)
Attleboro, Mass.

Charles H. Allen & Company
(JC 1896, 1904, 1915)
Attleboro, Mass.
Gold-filled jewelry. 1910 ad states "manufacturers of link buttons, fobs, hat pins, and 'Challenge' one-piece collar buttons. Offices in New York and Chicago."

C. H. ALLEN W. H. LAMB

C. H. ALLEN & CO.

Manufacturers of

Link Buttons, Fobs, Hat Pins, Etc.

also

"Challenge" one piece Collar-Buttons

Factory: ATTLEBORO, MASS.

New York Office: 180 Broadway

Chicago Office: 704 Heyworth Bldg.

The Clingtooth

Trade Mark

Allen, Smith, & Thurston Co.
(Succeeded by Allen, McNerney, & Co.)
(JC1905)
Attleboro, Mass.
Shell goods.

Attaya Bros. Inc.
(Founded 1948)
Attleboro, Mass.
Relgious and other medals.

TRADE
HAA&CO
MARK

(*Discontinued.*)

H. A. A. & CO.
H. A. ALLEN & CO.,

(*Discontinued.*)

H.A. Allen & Company
(c. 1909-1922)
Attleboro, Mass.
1910 ad states "makers of high grade medium priced links, hat pins, waist sets, fobs, scarf pins, novelties, offices in New York, Chicago, San Francisco."

H. A. ALLEN & CO.
Ingraham Bldg., Attleboro, Mass.
Makers of
High Grade Medium Priced

Links	**Hat Pins**	**Waist Sets**
Fobs	**Scarf Pins**	**Novelties**

New York Office: -	**194 BROADWAY**
Chicago: - - -	**103 STATE ST.**
San Francisco: - -	**704 MARKET ST.**

A. L. S. N.

Allison Manufacturing Company
(c. 1909)
Successors to Regnell, Bigney, and Company
Bracelets.

JOHN ANTHONY
Manufacturer of Chains
ATTLEBORO, MASS.
WOVEN WIRE CHAINS FROM 1-8 TO 1 1-4 INCHES WIDE
Also Manufacturer in
Gold, Silver, Plate, Brass, Steel and Aluminum
PADLOCKS, (Crandall Patent) MACHINE, EYE GLASS, CABLE and CURB CHAIN
LINKS OF ALL KINDS FOR MESH BAGS.
JUMP RINGS ALL SIZES.

John Anthony
(c. 1910)
Attleboro, Mass.
1910 ad states "manufacturer of chains-gold, silver, plate, brass, steel, and aluminum. Links of all kinds for mesh bags."

Attleboro Chain Company
(1909-1915)
Attleboro, Mass.
An advertisement of 1910 reads "Manufacturers of rolled plate jewelry-the Marathon line-chains, fobs, lockets, bracelets, pendants, La Vallieres, and the famous innergroup locket."

(On Toilet Ware.)
(Discontinued.)

ATTLEBORO CHAIN CO.,

(Novelties.)
(Discontinued.)

Attleboro Manufacturing Company
(JC 1898-1915, out of business by 1922)
Succeeded by Swank.
Attleboro, Mass.
An advertisement of 1910 reads "On account of the general excellence of its manufacture, are easily recognized as makers of the foremost line of ladies jewelry and novelties in America. We are patentees of the famous Adjustable 'COMMENT' bracelets. Our line of waist sets, brooches, cuff pins, hat pins, sash pins, jabot pins, belt buckles, clinchtite solderless cuff and collar pins, coin purses. If you are looking for novelties, come here. Silver Medal-Paris Exposition-Bronze Medal. Office in New York."

ATTLEBORO MANUFACTURING CO.

INCORPORATED

On account of the general excellence of its manufacture, are easily recognized as makers of the

Foremost Line of Ladies' Jewelry and Novelties in America

We are patentees of the famous Adjustable "COMMENT" Bracelets. Our line of Waist Sets. Brooches, Cuff Pins, Hat Pins, Sash Pins, Jabot Pins, Belt Buckles, Clinch-tite Solderless Cuff and Collar Pins, Coin Purses. If you are Looking for NOVELTIES, come here.

Silver Medal — PARIS EXPOSITION — Bronze Medal

NEW YORK: 9 Maiden Lane, WORKS: Attleboro, Mass.

Augat, Inc.
(c. 1931)
Attleboro, Mass.
Jewelry findings and ornaments.

BALFOUR
LGB

L.G. Balfour
(1913-present)
Attleboro, Mass.
Fraternity jewelry and class rings for high schools, colleges, and universities. Trophies and medals for sports. Moved operations to Texas.

Austin & Stone, Inc.
(JC 1909 & 1915)
Attleboro, Mass.
Gold chain. In 1910, offices in Chicago and SanFrancisco.

THE JOBBERS' HANDBOOK. 83

BALLOU MFG. CO.
ATTLEBORO, MASS.

Manufacturers of
LOCKETS, PENDANTS, GOLD FRONT BUTTONS and CUFF PINS

Also a line of Brooches, Buckles, Scarf, Hat, Veil and Sash Pins.

Quality and Price Unrivalled

WAIT FOR OUR SALESMAN WITH SAMPLES

Ballou Manufacturing Company
(c. 1910-1927)
Attleboro, Mass.
Gold, gold-filled and sterling silver jewelry, materials and findings. 1910 ad states "manufacturers of lockets, pendants, gold front buttons, buckles, scarf, hat, veil, and sash pins."

B. MFG. CO.

Baer & Wilde Company
(JC 1915, K 1922 & 1931)
Attleboro, Mass.
Kum-a-part cuff buttons, full dress sets, collar holders, tie and money clips, collar buttons and tie pins. 1910 ad states "makers of correct jewelry for men-link buttons, fobs, tie clips, scarf pins, and gents' combination sets. Office in New York."

THE BAER & WILDE CO.
Makers of
Correct Jewelry for Men

Link Buttons, Fobs, Tie Clips, Scarf Pins and
GENTS' COMBINATION SETS

Main Office and Factory: ATTLEBORO, MASS.

New York Office: 9-13 Maiden Lane
SAMPLES ONLY

C. & B.

B. B. & CO.

STERLING.

Barden, Blake & Co.
(c. 1896-1915, succeeded by Chapman & Barden; "out of business")
Attleboro, Mass.

B&H

BARDEN & HULL
Makers of
Solid Gold Jewelry
.. SCARF PINS ..
Brooches, Ear Drops, Gold Mounted Combs and Barrettes. High Grade Rolled Plated Fobs and Lapel Chain. Electric Soldered Single Link Neck Chain.
ATTLEBORO, - - MASS.
A. B. COPELAND, 9 Maiden Lane, New York.
F. R. SHERIDAN, 503 Chronicle Bldg., San Francisco.

Barden & Hull
(c. 1910-1915)
Attleboro Falls, Mass.
1910 ad states "makers of solid gold jewelry-scarf pins-brooches, ear drops, gold mounted combs and barrettes. High grade rolled plated fobs and lapel chain. Electric soldered single link chain. Offices in New York and San Francisco."

H.F.B.&Co.

H. F. B. & CO.
(Discontinued.)

H. F. B.

H. F. Barrows & Company
(Founded 1851- JCK 1934)
North Attleboro, Mass.
Sterling silver, gold filled, and gold jewelry.

(On Swivels.)

B. & B.
(Lockets and Tissues for Ladies' Chains.)

BATES & BACON
(On Swivels.)

(On Stationery and Tissues.)

Bates & Bacon
(c. 1856-)
Attleboro, Mass.
The production of watch cases was added in 1882, the company being the first in Attleboro to offer this line.
1910 ad states "established 1856, manufacturers of high grade gold filled chains, lockets, and bracelets-chains best safety fob, gents vest dielans, outing, lorgnettes, matinee 54 & 60 in., secret locket 13-22-22 in., eye glass, bracelets, lockets, chatelaines, crosses, fob clasps, and attach chains.

Bates & Klinke
(c. 1919-present)
Attleboro, Mass.
Convention and souvenir jewelry. Moved to Johnston, Rhode Island

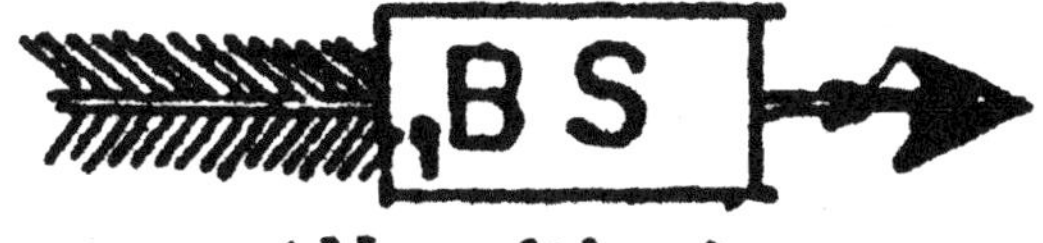

(Novelties.)

Bay State Silver Company
(JC 1915)
North Attleboro, Mass.

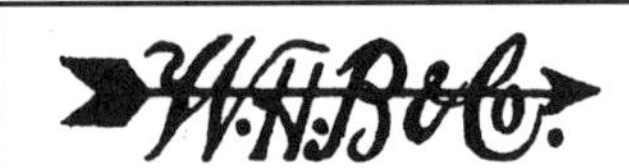

B. & CO.
W. H. B. & CO.

WM. H. BELL & CO.

Manufacturing Jewelers,

ATTLEBORO FALLS, MASS.

SPECIALTIES:

Gold Filled and Plated Vest and Fob Chains of every description; also Eye Glass Chains and Chain Trimmings. Lorgnettes with Solid Gold and Plated Slides

Bead Neck, Secret Locket Chains and Neck Chains

Wm. H. Bell & Co.
(c. 1910)
Attleboro Falls, Mass.
1910 ad reads "manufacturing jewelers-specialties-gold-filled and plated vest and fob chains of every description: also eyeglass chains and chain trimmings. Lorgnettes with solid gold or plated slides. Bead neck, secret locket chains, and neck chains.

B. & S.

Bennett & Sawyer Company
(JC 1915)
Attleboro, Mass.

BISHOP FASHIONED

Bishop Company
(c. 1950)
North Attleboro, Mass.

(Discontinued.)

(Nethersole Bracel

STERLING SILVER FLATWARE

Alden	Nassau
Camden	Sweet Briar
Cherry Blossom	Verona
Empress	York
Helena	

(Discontinued.)

Azalea	De Coverly
Claremont	Narcissus
Daisy	

STERLING SILVER TOILETWARE

Acanthus	Plain
Berkshire	Rajah
Engraved	Regent
Essex.	Rex
Franklin	Royal
Garland	Salem
Hampden	Stuart
Norfolk	Sultan

R. Blackinton & Company
(1862-1978)
North Attleboro, Mass.
Sterling and 14K gold novelties, including guilloche sterling enamel dresser ware, flatware, and hollowware.

BIGNEY
CAMILLE
LADY CAMILLE
LADY SILVIA
S.O.B.
MIRROR FINISH

THE JOBBERS' HANDBOOK. 179

S. O. BIGNEY & CO.

ATTLEBORO, MASS.

HIGH GRADE FILLED CHAINS
WARRANTED
TRADE MARK

Manufacturers of
The Mirror Finish
HIGH GRADE
Gold - Filled
CHAINS

Lockets, Bracelets and all kinds of Novelties in Jewelry
New York Office: 3 Maiden Lane

S.O. Bigney & Company
(c. 1903-1943)
1910 ad states manufacturers of the mirror finish-high grade-gold filled chains, lockets, bracelets, and all kinds of novelties in jewelry.

BLACKINTON

GOLD-X **SILV-X**

HI-GLO

V.H. Blackinton & Company
(K 1922 & 1931)
Attleboro Falls, Mass.
Emblems, badges, military insignia.

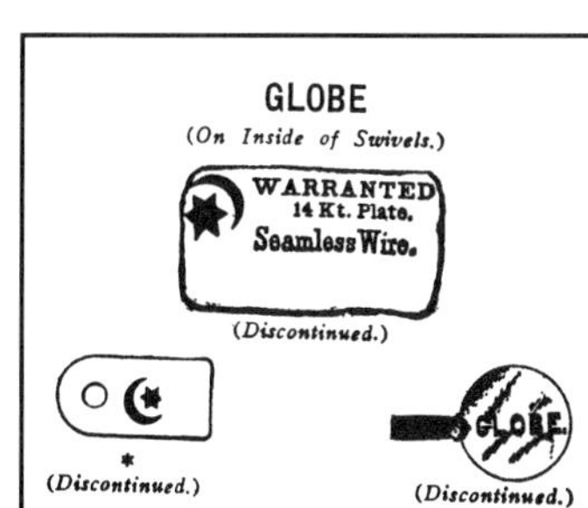

PIONEER
(Discontinued.)

BLACKINTON
OLD RELIABLE
W. & S. B. ★

W & S. Blackinton
(c. 1865-present)
Attleboro, Mass.
Gold and gold-filled jewelry.
Moved to East Boston, Mass.

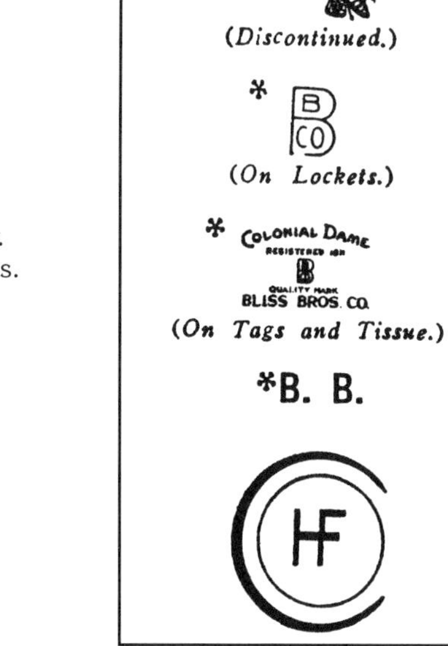

Bliss Bros. Company
(c. 1873-1950)
Attleboro, Mass.
1910 ad reads "Manufacturing jewelers-lockets, bracelets, hat pins, silk fobs, and charms, link buttons, scarf pins, waist sets. Offices in New York and Chicago."

STERLING SILVER TOILETWARE

Applied Bead	Manhattan
Berkeley	Monticello
Brentford	Plymouth
Kenwood	Thread
Lady Mansfield	York

STERLINE

James E. Blake Company
(c. 1880-1936)
Attleboro, Mass.
On January 31, 1905, received a patent for the manufacture of sterling silver and silver inlaid with 14K gold cigarette and vanity cases, match boxes, napkin rings, men's belt buckles, and pocket knives."Sterline" jewelry and dresser ware. Sterling enamel compacts. From 1910 ad "sterling silver toilet ware & novelties, sandwich plates, compotes, vases, etc."

THE JOBBERS' HANDBOOK. 139

Established 1873. E. M. BLISS, Pres.
Incorporated 1901. H. C. BLISS, Treas.

B.B.

BLISS BROS. COMPANY,

MANUFACTURING JEWELERS

Lockets, Bracelets, Hat Pins, Silk Fobs & Charms, Link Buttons, Scarf Pins, Waist Sets.

Factory : ATTLEBORO, MASS.

NEW YORK: 15-17-19 Maiden Lane. E. M. COE, Representative.
CHICAGO: 1110 Heyworth Bldg. C. P. CRANE, Representative.

A. H. B. & CO.

(On Tags and Tissues.)

A.H. Bliss Co.
(Established 1880, JC 1904 & 1915, K 1931)
North Attleboro, Mass.
1910 ad states "Manufacturers of chains, including vests, guards, dickens, necks, ponys, festoons, fobs, and eye-glass chains. Also made bracelets 'Secret Lock' and expansion. Chains include solid gold, seamless gold filled, rolled gold plate, sterling silver. Chains and findings for the manufacturer. Also padlocks."

A. H. BLISS CO.
Established 1880
.... Manufacturers of ....

Our Line Includes the Following:
VESTS, GUARDS, DICKENS, NECKS, PONYS, FESTOONS, FOBS, EYE-GLASS CHAINS.

Trade

Mark

BRACELETS
Secret Lock, and Expansion
Our New Fobs are Up-to-date

FACTORY: NORTH ATTLEBORO, MASS.
NEW YORK: 9 Maiden Lane

Bodman Bros.
(JC 1915)
Attleboro, Mass.

W. F. B. & CO.

Wm. F. Briggs & Company
(Out of business before 1915)
Attleboro Falls, Mass.

JOHN P. BONNETT & SON.

Established 1879

**ELECTRO-METALLURGISTS
and
COLORERS OF JEWELRY**

Plain, Fancy, Decorative and Ornamental Colorings
of Every Description

North Attleboro, Massachusetts

John P. Bonnett & Son
(c. 1910)
North Attleboro, Mass.
Electro-metallurgist

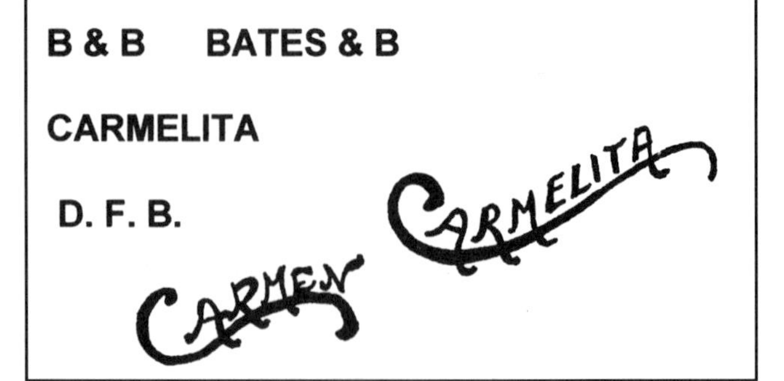

Briggs, Bates, and Bacon Company
(Successors to D.F. Briggs Company 1922, JC-K 1950)
Carmen and Carmelita expansion bracelets. Neck chains.

Boss & Baldwin Company
(JC 1909)
Attleboro Falls, Mass.

Bristol Silver Company
(JC 1915)
Attleboro, Mass.

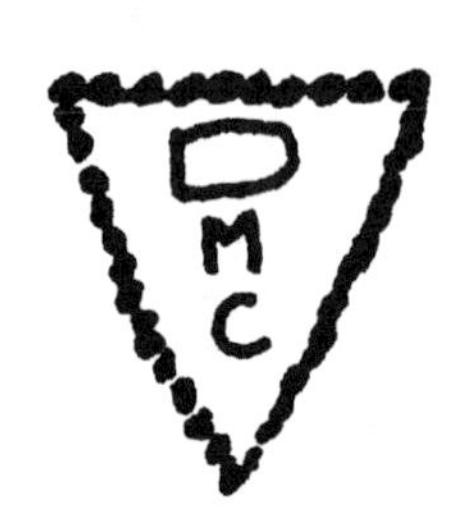

**D. F. B. CO.
CARMEN**

D.F. Briggs Company
(c. 1890-1922, succeeded by Briggs, Bates, and Bacon)
Attleboro, Mass.
An advertisement of 1910 reads "The best rolled gold plate chain in the world for your money. Every chain warranted. Manufacturers of gold filled chains, fobs, locket chains, necks, hat pins, veil pins, chateleine pins, bracelets, and charms. Offices in New York, Chicago, St. Louis, and London." Makers of the Carmen and Carmelita expansion bracelets. Enamel compacts.

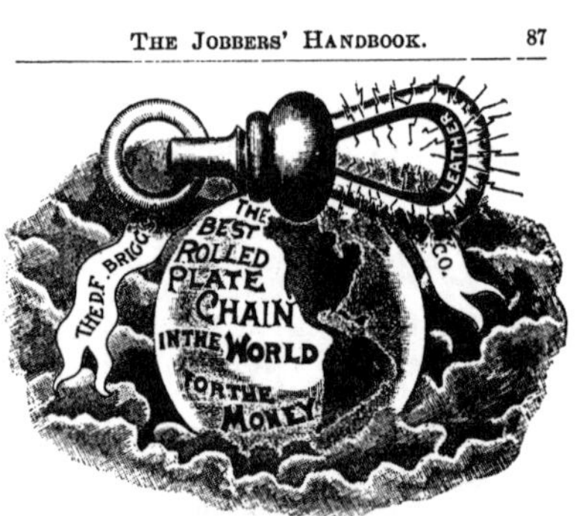

EVERY CHAIN WARRANTED
The D. F. BRIGGS CO.,
ATTLEBORO, MASS.
Manufacturers of GOLD FILLED CHAINS, FOBS, LOCKET CHAINS, NECKS, HAT PINS, VEIL PINS, CHAT. PINS, BRACELETS and CHARMS.

NEW YORK OFFICE: 180 Broadway
CHICAGO OFFICE: Heyworth Building
ST. LOUIS OFFICE: Victoria Building
LONDON OFFICE: 62 Hatton Garden

Geo. L. Brown Company
(JC 1904 & 1915, K
1922)
Jewelry and scenic
enamel compacts.

C. & B.

Chapman & Barden
(JC 1915 "out of business")
Attleboro, Mass.

Chartley Jewelry
Mfg. Co.
(JC 1915)
Chartley, Mass.

C. T. & CO.

Cheever, Tweedy &
Company
(Succeeded J.G. Cheever
& Company 1915, listed
1977)
North Attleboro, Mass.

THE JOBBERS' HANDBOOK. 195

Established 1869

A. BUSHEE & CO.

Makers of Gold Filled and Plated

:: JEWELRY ::

ATTLEBORO, -:- MASS.

Hat Pins, Sash Pins, Pendants, Brooches, Collar, Veil and Bar Pins, Pin Sets and Novelties, Cuff Links, Scarf Pins, Link and Scarf Sets, Tie Clasps, Collar Buttons, Etc. " Etruscan Wire Work a Specialty."

Samples and Prices to Responsible Jobbers

Chicago, Ill. Heyworth Bldg. **San Francisco, Cal. 717 Market St.**

A. Bushee & Co.
(founded 1869)
An advertisement of 1910 reads "Established 1869-makers of fine gold filled and plated jewelry-hat pins, sash pins, cuff links, scarf pins, pendants, brooches, link and scarf sets; collar, veil and bar pins, tie clasps, collar buttons, pin sets and novelties; Etruscan wire work a specialty. Offices Chicago and San Francisco."

J. G. C. & CO.

J.G. Cheever & Company
(c. 1880-1915, succeeded by Cheever, Tweedy & Company)
North Attleboro, Massachusetts

THE JOBBERS' HANDBOOK. 355

Providence. Attleboro. North Attleboro.

GEORGE L. CLAFLIN CO.

Jewelers' Supplies

LACQUER—For all Grades of Work
Agents of the Egyptian Lacquer Mfg. Co.
F. F. ANTI-OXIDIZING COMPOUND,
Superior to Boracic Acid,
Try a 5 pound package.
LEATHER BELTING—All listed sizes up to 6 inch in stock.
KALYE—In 25 pound cans. Removes dirt and grease—does not tarnish. N. E. Agents.
CYANIDE POTASSIUM—Red Label Brand, 5, 10, 25 and 112 pound cans
CROCUS, G. L. C. BRAND—The Best Value.
ACIDS, AMMONIA, CHEMICALS and CHEMICAL GLASSWARE, POLISHING MATERIALS, RUBBER FINGER COTS, C. H. BRUSHES, ETC., ETC.

TELEPHONES:
Providence Branch Exchange, 3835 Un., connecting all departments.
Attleboro Telephone, 48-L.
North Attleboro Telephone, 141-L.
REPRESENTED BY C. A. FISHER.

George L. Claflin
(c. 1910)
Attleboro, North Attleboro, Mass.

C. Q. & R.
(On Swivels.)

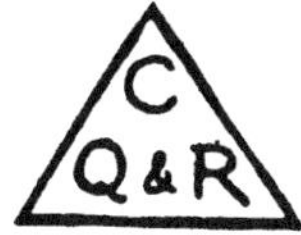

Carter, Qvarnstrom & Remington

Manufacturers of

GOLD and HIGH GRADE ROLLED PLATED

Lockets, Charms, Miniature Brooches, Fob and Neck Chains

Bigney Bldg. County St. ATTLEBORO, MASS.

NEW YORK OFFICE:

Room 1203 **9-11-13 Maiden Lan**

Carter, Qvarnstrom and Remington
(JC 1904 & 1915, K 1922)
An advertisement of 1910 reads "Gold and high grade rolled gold plated lockets, charms, miniature brooches, fob and neck chains." Sterling enamel purse perfumes.

Clark

W. G. C.

(Discontinued.)

W.G. Clark & Company
(JC 1904 & 1915, K 1922)
North Attleboro, Mass.

F. M. & J. L. Cobb
(JC 1915)
Mansfield, Mass.

C. B. & H.

STERLING.

C. B. & H.

(Discontinued.)

Codding & Heilborn Company
(1862-1918)
Bracelets, signet rings, secret catches, springs and bangles, sash pins, hatpins, buckles, lorgnette chains and brooches. Sterling cigarette cases.

STERLING.

D. D. C.

D. D. Codding
(JC 1915)
North Attleboro, Mass.

College Seal & Crest Company
(c. 1875-1965)
High school and college jewelry, stone set or enameled in authentic school colors.

ANNE CECILE

ANNE ELAINE

ANNE LOUISE

Clap, Harvey & Company
(Succeeded Daggett & Clap Co. prior to 1915)
Gold and gold-filled jewelry for men and women, watch bracelets, baby bracelets, pins and cuff pins.

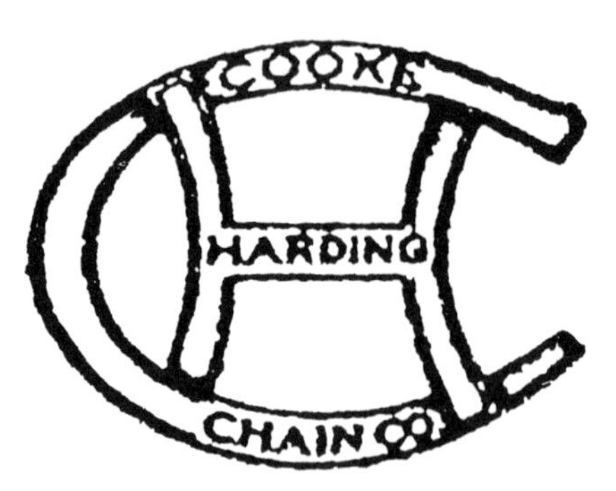

TRADE MARK

Cooke, Harding Chain Company
(c. 1912-)
Later purchased by Sweet Manufacturing Co.

Craft, Inc.
(1950-)
South Attleboro, Mass.
Jewelry and small metal stampings.

Creed Rosary Co., Inc.
(1948-present)
North Attleboro, Mass.
Religious jewelry, including rosaries, medals, crosses, key chains and cufflinks in gold and gold filled.
Sterling enamel charms with religious themes.

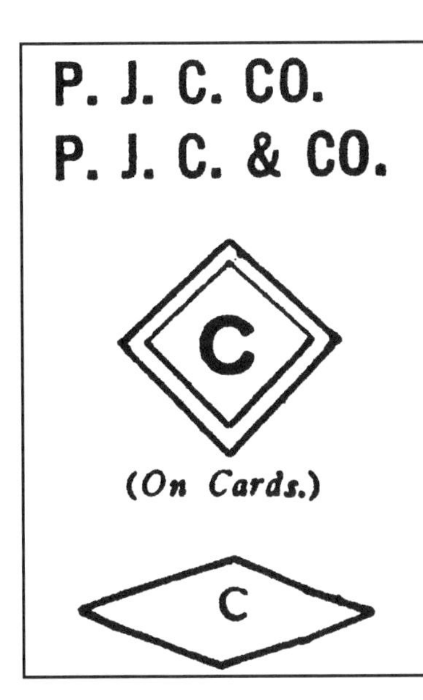

P.J. Cummings Company, Inc.
(JC 1910, "Out of business" 1915)
Gold chains.

H.H. Curtis Company
(1891-1915)
North Attleboro, Mass.
Jewelry, table and flatware.

F. H. Cutler & Company
(JC 1904 & 1915)
North Attleboro, Mass.
Jewelry manufacturer.

Cutler & Granberry
(JC 1896, became J.A. & S.W. Granberry before 1904)
North Attleboro, Mass.

D. & C.

Daggett & Clap Co.
(JC 1915, succeeded by Clap, Harvey & Co.))
Attleboro, Mass.
Gold filled jewelry.

Julian C. Daniels
(JC 1915)
Attleboro, Mass.

The D. & D. CHAIN CO.
Makers of
Machine Chain, Curb Chain
Rope Chain, French Curb
Chain, Etc.
in Platinum, Gold, Silver, Plate and Brass
Winding and Sawing of Links of all Kinds
23 1-2 County St., -:- ATTLEBORO, MASS.

D & D Chain Co.
(c. 1910)
Attleboro, Mass.
An advertisement of 1910 reads "If its chain we make it. Makers of machine chain, curb chain, rope chain, French curb chain, etc. in platinum, gold, silver, plate, and brass-winding and sawing of links of all kinds."

S. D.
CHAMPION

Saloman Davidson
(JC 1915 "Out of business")
Attleboro, Mass.

G. A. D. CO.
G. A. D. & CO.
(Discontinued Jan. 1, 1903.)

G.A. Dean Company
(c. 1903, succeeded by Bates & Bacon 1904)
Attleboro, Mass.

THE JOBBERS HANDBOOK. 211

DOBRA BROS. CO.
ELECTROPLATERS
Always Something New
Plain, Fancy and Ornamental Coloring
NICKEL, BRASS and COPPER PLATING
and POLISHING
36 Railroad St., ATTLEBORO, MASS.
Telephone Connection

Dobra Bros. Co.
(c. 1910)
Attleboro, Mass.

(On Tissues.)

Doran, Bagnall, & Company
(c. 1870, JC 1915, K. 1922, succeeded Young, Bagnall, & Company before 1905)
North Attleboro, Mass.
An advertisement of 1910 reads "Established 1870-manufacturers of fine rolled gold plate and sterling silver-chains, bracelets, and novelties. See our new line of more than 60 patterns of sterling silver la vallieres and bracelets and high grade rosaries. Offices New York, Chicago, San Francisco."

PERFECT
NO FUSS
S. D. & CO.

Samuel Dosick & Company
(K 1922)
Attleboro, Mass.
Men's jewelry.

14K D.
O. M. D.
NONPAREIL
PREMIER
ZENITH NONPAREIL
(chains)

PREMIER
(Discontinued.)

***NONPAREIL**
O. M. D.
***ZENITH**

O.M. Draper Company
(c. 1862, later a division of Le Stage Manufacturing Company)
North Attleboro, Mass.

C.H. Eden Company
(c. 1906-1931)
"Princess Alice" adjustable bracelet and fancy combs.

MINERVA

E. C. CO.

Electric Chain Company
(c. 1900-1931)
Chains and findings, adjustable bracelets and other jewelry.

Elliot & Douglass Manufacturing Company
(JC 1915, K 1922)
Rolled gold plate and gold front jewelry.

E. L. & B.

Ellis, Livsey & Brown
(JC 1904 "Out of business")
Attleboro, Mass.
Plated neck chains.

EVANS

Evans Case Company
(c. 1922-1960)
North Attleboro, Mass.
Compacts, purses, perfumes, lighters, cases. Specialized in enameling. Later a division of Hilsingor Corp., Plainville, Mass.

Excel Rosary & Novelty Company, Inc.
(K 1931, JC-K 1965)
North Attleboro, Mass.

E.A Fargo Company
(c. 1903)
Attleboro, Mass.
"Silver-aluminum" novelties.

FMI **JCF**

J & C. Ferrara Company
North Attleboro, Mass.
Jewelry in gold, sterling, vermeil and pewter.

Fillkwik Company
(c. 1920-1936)
Attleboro, Mass.
Became Shields, Inc.

F. M. CO.

F. M. C.
(Canada)

F. M. Co.
(Domestic & international)

THE JOBBERS' HANDBOOK. 93

ESTABLISHED 1890
FINBERG MFG. CO.
MAKERS OF
FINBERG'S FAULTLESS FOBS
Bracelets, Lockets, Necks, Pendants
Finberg's Faultless Fobs, Lockets & Bracelets
THE "F. M. CO." LINE
Attleboro, Mass.

Finberg Manufacturing Company
(c. 1890-JC-K 1934)
An advertisement of 1910 reads "Established 1890-makers of Finberg's faultless fobs, bracelets, lockets, necks, pendants."

Findex Corp.
Attleboro, Mass.
Wireforming and
earwires.

Flagg Manufacturing Company
(c. 1931)
Attleboro, Mass.
Novelty jewelry and specialties.

J. M. F. CO.

J. M. F. & CO.

FISHER

LOVEBRIGHT

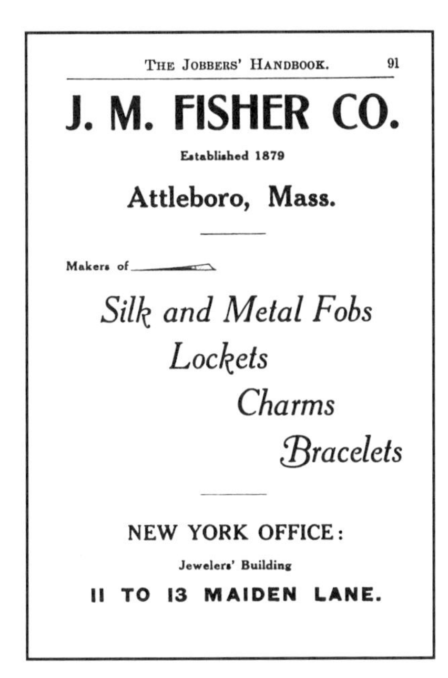
THE JOBBERS' HANDBOOK. 91

J. M. FISHER CO.

Established 1879

Attleboro, Mass.

Makers of

Silk and Metal Fobs
Lockets
Charms
Bracelets

NEW YORK OFFICE:
Jewelers' Building
11 TO 13 MAIDEN LANE.

J.M. Fisher Company
(1879-present, Division of LeStage Manufacturing Company)
An advertisement of 1910 reads "Makers of silk and metal fobs, lockets, charms, bracelets." Compacts and novelties. Gold, gold-filled and sterling jewelry.

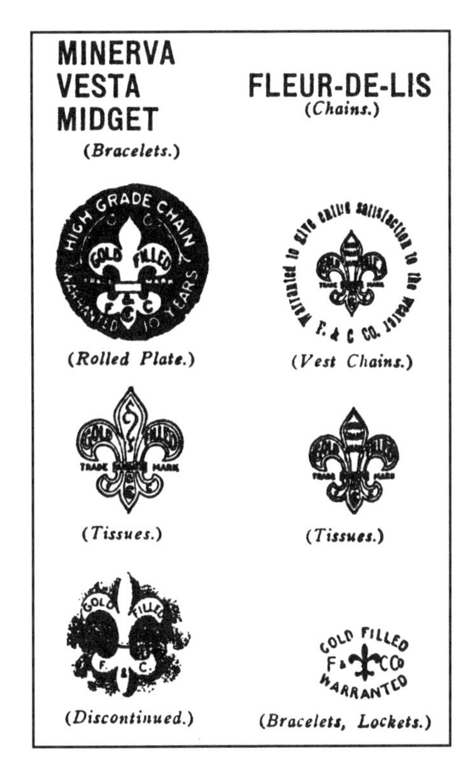

MINERVA
VESTA
MIDGET
(bracelets)

FLEUR-DE-LIS
(chains)

FONTNEAU & COOK CO.
Manufacturers of
FINE ROLLED PLATE AND
GOLD FILLED JEWELRY

Bracelets
Pendants
Fobs
Chains
and
Lockets

Swellest
Line
on
the
Road
to-day

Factory: ATTLEBORO, MASS.
New York Office: 15 Maiden Lane
Chicago Office: 1203 Heyworth Bldg.

Fontneau & Cook
(JC 1904 & 1909)
An advertisement of 1910 reads "Manufacturers of fine rolled plate and gold filled jewelry. Bracelets, pendants, fobs, chains and lockets-swellest line on the road today. Offices in New York and Chicago." Makers of the "Midget Adjustable Baby Bracelet."

THE JOBBERS' HANDBOOK. 217

W. N. FISHER & CO.

ESTABLISHED 1877

Manufacturing Jewelers

ATTLEBORO FALLS,
MASS.

Chain Trimmings
In Gold, Rolled Plate and Sterling Silver

SPRING SWIVELS
SPRING RINGS

Telephone:
8-J NO. ATTLEBORO.

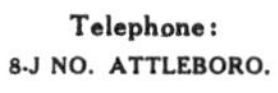

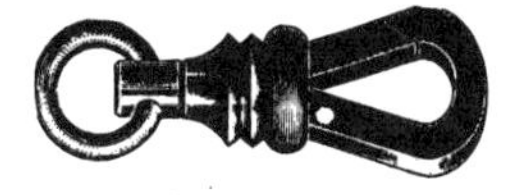

W.N. Fisher Company
(c. 1877-, JC 1915, K 1922 & 1931)
Attleboro Falls, Mass.
An advertisement of 1910 reads "Established 1877-manufacturing jewelers-chain trimmings in gold, rolled plate, and sterling silver. Spring swivels and spring rings."

E. I. F. & CO.

THE CENTURY.

CENTURY

E.I. Franklin & Company
(JC 1904 & 1915, K 1922)
North Attleboro, Mass.
Gold-filled and sterling silver jewelry, including enamel.

B. S. F. & CO.
(Discontinued.)
B. S. F. CO.
(Discontinued.)

J.J. and B.S. Freeman Company
(founded 1847)
Attleboro Falls, Mass.
Jewelry manufacturers. The first to manufacture curbed chains of rolled gold in the United States with machinery invented by J.J. Freeman.

T.G. Frothingham & Company
(JC 1915, K 1922 & 1931)
Gold jewelry and ring mountings.

F.I. Gorton
(K 1931)
North Attleboro, Mass.
Emblems.

(Discontinued.)

B. & S.

Hansen-Bennett Company
(JC 1909) Succeeded by Bennett & Sawyer Company before 1915.

H. & S.
(Discontinued.)

(Discontinued.)

Hayward & Sweet
(c. 1887-1904) Succeeded by Walter E. Hayward & Company.

F

TRADE F MARK.

L.E. Freeman Company
(JC 1915, K 1922)
North Attleboro, Mass.
Gold jewelry, cigarette holders, and knives.

SP.

GENCO
AURORA
RECORD

General Chain Company
(JC-K 1943, 1977)
North Attleboro, Mass.

General Findings
(1914-present) Division of Leach & Garner Company
Attleboro, Mass.
Spring rings, swivels, and related findings for the jewelry and chain industry.

Chas. K. Grouse Company
(JC 1909 & 1915, K 1922 & 1931, JC-K 1943 & 1950)
North Attleboro, Mass.
Badges, emblems, class rings and pins.

MONA LISA

D. A. HART CO.
ATTLEBORO - MASSACHUSETTS
Makers of High Graue
Hat Pins, Sash Pins,
Pendants, White Stone Goods
37 Maiden Lane, New York — 910 Heyworth Bldg. Chicago
Address all Correspondence to Attleboro

D.A Hart Company
(C. 1910, JC 1915)
An advertisement of 1910 reads "Makers of high grade hat pins, sash pins, pendants, white stone goods."

B & H

Freeman, Barden & Hull Company
(succeeded Barden & Hull Company before 1915)
Attleboro, Mass.

STERLING G

F. S. Gilbert
(JC 1915 "out of business")
North Attleboro, Mass.

Guyot Brothers Company
(c. 1904-)
Attleboro, Mass.
Jewelry finding and custom stampings.

French & Franklin
(JC 1896 & 1904, JC 1915 "Out of business")
North Attleboro, Mass.
Sterling silver flatware and novelty items.

THE JOBBERS' HANDBOOK. 95
E. D. GILMORE & CO.
Manufacturers of
JEWELRY
Rings, Studs, Scarf Pins, Brooches, Drops, Hat Pins
In Gold and Solid Gold Mountings
ATTLEBORO, :: :: MASS.

E.D. Gilmore & Company
(C. 1910)
Attleboro, Mass.
An advertisement of 1910 reads "Manufacturers of jewelry-rings, studs, scarf pins, brooches, drops, hat pins-in gold and silver mountings."

W. E. H.
(On Swivels.)
(Discontinued)
HAYWARD
W. E. H. CO.
(On Swivels.)

Walter E. Hayward & Company
(1851-) Established as Thompson, Hayward & Company, also called Hayward & Briggs, Hayward & Sweet, C.E. Hayward Company, became Walter E. Hayward & Company before 1904, and Amtel Arts in 1971.
Attleboro, Mass.
E. Providence, R.I.
Gold, gold filled, and sterling jewelry.

J.T. Healy
(K 1922 & 1931, JC-K 1977)
Attleboro, Mass.
Metal beads and tubing products.

HOBSON

J.H. Hobson Company
(JC-K 1950)
North Attleboro, Mass.

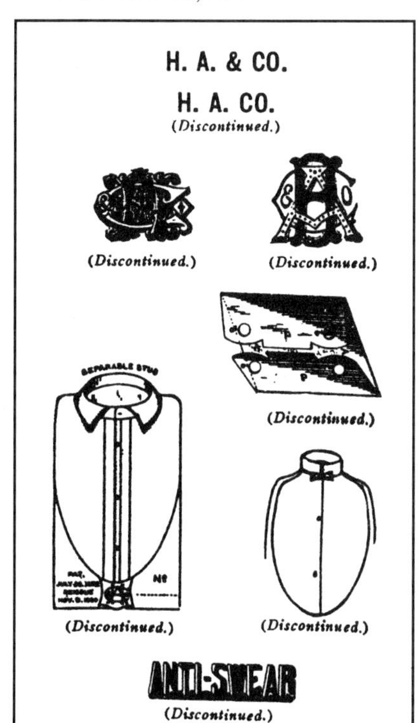

G. C. H. & CO.

G.C. Hudson & Company
(JC 1915, K 1922 & 1931)
North Attleboro, Mass.
Jewelry novelties.

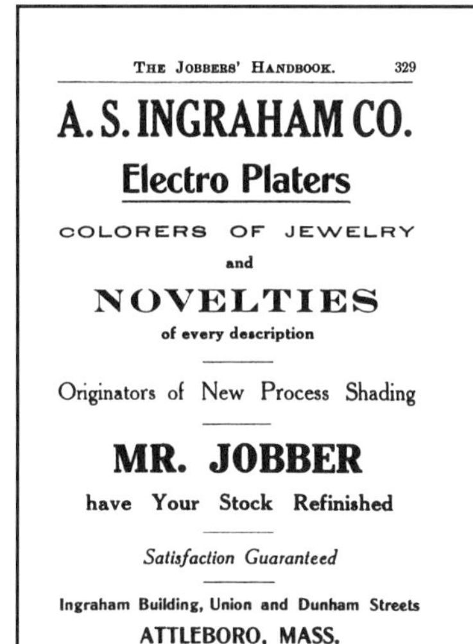

THE JOBBERS' HANDBOOK. 329

A. S. INGRAHAM CO.

Electro Platers

COLORERS OF JEWELRY
and
NOVELTIES
of every description

Originators of New Process Shading

MR. JOBBER
have Your Stock Refinished

Satisfaction Guaranteed

Ingraham Building, Union and Dunham Streets
ATTLEBORO, MASS.

A.S. Ingraham Co.
Attleboro, Mass.
Electroplating.

STERLING.

J.T. Inman & Company
(founded 1892-) In 1964, purchased by Whiting & Davis. When the Watson Company went out of business in 1955, the Inman Company bought some of their dies.

Horton, Angell & Company
(c. 1870-JC 1915 as Horton, Angell Co. Inc.)
Attleboro, Mass.
Jewelry manufacturer and metal supplier of karat gold, sterling silver, and gold-filled qualities.

Jeweled Cross Company, Inc.
(c. 1923-present)
Attleboro and North Attleboro, Mass.
Manufacturer of crucifixes.

J &CNCo
Gold Top

Jewelry & Cutlery Novelty Company
(JC-K 1950-1965)
North Attleboro & Cambridge, Massachusetts

J & L Tool & Findings Company
(c. 1912-)
Attleboro, Mass.
Jewelry findings.

Kennedy & Company
(K 1922)
North Attleboro, Mass.
Jewelry.

Keystone Jewelry Manufacturing Company
(JC 1905, JC 1915 "Out of business")
Attleboro, Mass.

THE JOBBERS' HANDBOOK. 203

J. R. KILBURN GLASS CO., Inc.
MANUFACTURERS OF
GLASS STONE
For Jewelers' and Silversmiths' Use

Imitation Stone of every variety
54 Union St., ATTLEBORO, MASS.

J.B. Kilburn Glass Co., Inc.
Attleboro, Mass.
Imitation Stone

Krew Incorporated
(c. 1950-)
Attleboro, Mass.
Military insignia and uniform accessories.

Lamb Manufacturing Company
(JC 1915)
Attleboro, Mass.

THE JOBBERS' HANDBOOK. 11

LEACH & GARNER COMPANY

Manufacturers of

Seamless Wire

Gold, Silver, Gold Filled, Plain and Fancy.

Seamless Tubing

Gold, Silver, Gold Filled and Brass in Ovals, Squares and Fancy Shapes.

Rolled Gold Plate

Gold, Silver and Silver Plate.

CHANNEL TUBING for bag frames, Plain and Fancy.
FANCY WIRES in all Metals.
GOLD and SILVER Anodes.
GOLD, SILVER AND BRASS SOLDERS. (Sheet, Wire, Cut and Burred.)

LEACH & GARNER BUILDING

ATTLEBORO, MASS., U. S. A.

Leach & Garner Company
Attleboro, Mass.
(1899-present)
Gold and gold-filled jewelry.

L. & M.

L. M. & CO.
(Discontinued.)

(On Tissues.)

Leach & Miller Company
Attleboro, Mass.
(JC 1904-1915, K 1922& 1931)
Sterling and gold-filled jewelry.

Leavens Manufacturing
Company, Inc.
(c. 1948-present)
Attleboro, Mass.
Emblems, badges,
medallions, watch fobs.

Le Stage Manufacturing
Company
(1863-present)
North Attleboro, Mass.
Gold and gold filled jewelry.

L. B. & CO.

LINCOLN, BACON & CO.,
Succeeded by
SCOFIELD, MELCHER & SCOFIELD,
PLAINVILLE, MASS.

The Lenau Co.
(JC 1915 "out of business")
Attleboro Falls, Mass.

L. B. & CO.
LINCOLN, BACON & CO.,
Succeeded by
SCOFIELD, MELCHER & SCOFIELD,
PLAINVILLE, MASS.

Lincoln, Bacon, and Company
(JC 1904)
Plainville, Mass.

Lindol Manufacturing Company
(K 1931)
Attleboro, Mass.
Gold rings, brooches, scarf and bar pins.

A. L. L. CO.

A.L. Lindroth Company
(K 1922 & 1931, JC-K 1943 & 1950)
North Attleboro, Mass.
Ladies and men's jewelry.

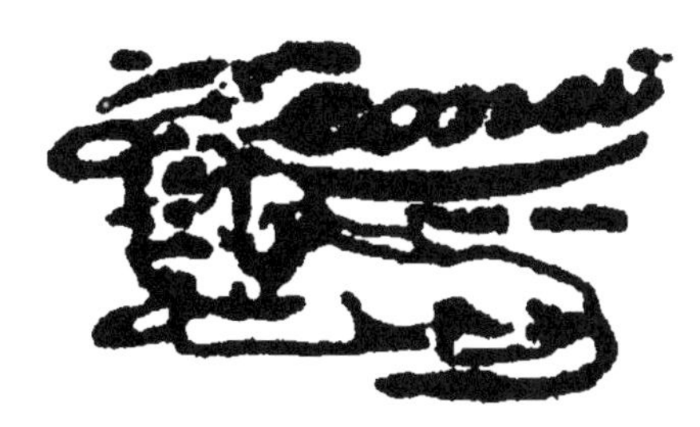

C.D. Lyons & Company
(JC 1909 & 1915, K 1922 & 1931)
Separable buttons.

R. B. M.
(Discontinued.)
BEATRICE
ALVURTES

THE JOBBERS' HANDBOOK. 219

LOCKETS - CUFF PINS AND LINKS

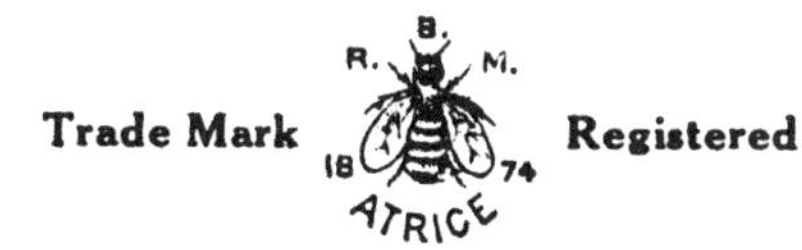

Manufacturing Jewelers

R. B. MACDONALD & CO.
Attleboro, Mass.

R.B. MacDonald & Company, Inc.
(c. 1874-1922)
Attleboro, Mass.
An advertisement of 1910 reads "Lockets, cuff pins and links-manufacturing jewelers." Gold-filled jewelry.

MG & Co.

MacKendrick Guyot & Company
(K 1922)
Attleboro, Mass.

M. & H.

THE JOBBERS' HANDBOOK. 235

MANDALIAN & HAWKINS
Manufacturing Jewelers
NORTH ATTLEBORO, MASS.
Extensive Line of
GERMAN SILVER and STERLING CHAIN BAGS and COAT-OF-MAIL PURSES
We carry a large variety to suit all demands.
ALSO POPULAR LINE OF BELT PINS.

Mandalian & Hawkins
(c. 1910-1921) Succeeded by Mandalian Manufacturing Company before 1922.
North Attleboro, Mass.
An advertisement from 1910 reads "Manufacturing Jewelers-German Silver and Sterling Chain Bags and coat-of-mail purses. We carry a large variety to suit all needs. Also popular line of belt pins."

MARSH

THE JOBBERS' HANDBOOK. 103

C. A. MARSH & CO.
The Quality Line
High Grade, Seamless, Gold Filled
CHAINS
Lockets, Bracelets
Pendants

Office and Factory:
Attleboro, Mass.

EDWARD N. COOK PLATE COMPANY
Manufacturers of
SEAMLESS WIRE AND TUBING
in all sizes and qualities
GOLD AND SILVER PLATE
Round, Square, Flat and Fancy Wire
Aluminum Solder, Anodes for Plating
RING TAPERS
METCALF BLD'G.
144 PINE STREET,
PROVIDENCE, R. I.

C.A. Marsh & Company
(JC 1904 & 1915, K 1922)
Attleboro, Mass.
An advertisement from 1910 reads "The quality line-high grade, seamless, gold filled chains, lockets, bracelets, and pendants." Buttons, lorgnettes, fobs, vest chains.
Sterling jewelry, including enamel.

M. B. & E.

Maintien Bros., Inc.
(JC 1915 Maintien Bros. & Elliot, K 1922 & 1931)
Plated and sterling jewelry.

THE JOBBERS' HANDBOOK. 259

ESTABLISHED 1856.

BATES & BACON
ATTLEBORO, - - MASS.
Manufacturers of High Grade
Gold Filled Chains, Lockets and Bracelets

CHAINS
Best Safety Fob
Gent's Vest
Dielans
Outing
Lorgnettes
Matinee 54 and 60 in.
Secret Locket 13-18-22 in.
Eye Glass

Bracelets
Lockets
Chatelaines
Crosses
Fob Clasps and Attach Chains

New York: 9 Maiden Lane.
L 2106
Chicago: 103 State Street.

D.E. Makepeace
Attleboro, Mass.
Gold and silver, rolled plate, and wire.

DEBUTANTE

Mandalian Manufacturing Company
(c. 1922-1935)
North Attleboro, Mass.
Sterling and nickel silver mesh bags and frames.

Martha Manufacturing Company
(K 1931)
Attleboro, Mass.
Mens and ladies jewelry in gold-filled and sterling silver.
Necklaces, bracelets, buckles, and buckle sets.

TEEN KRAFT

Kiddie Kraft
SUB-DEB
LADYETTE
PERM-A-PLATE
DOLLY MADISON

Marathon Company
(c. 1897 Attleboro Chain Co., 1922-present as Marathon)
Attleboro, Mass.
Gold, gold filled and sterling jewelry. Compacts, cases, and lighters.

Martha & Carpenter
(K 1922)
North Attleboro, Mass.

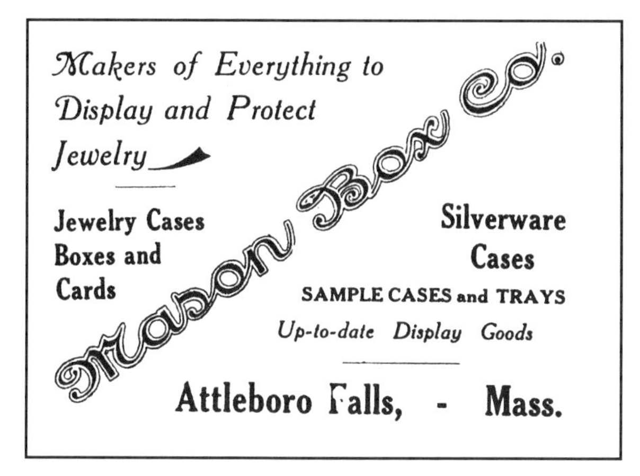

Mason Box Co.
Attleboro Falls, Mass.
Jewelry Cases, Boxes and Cards

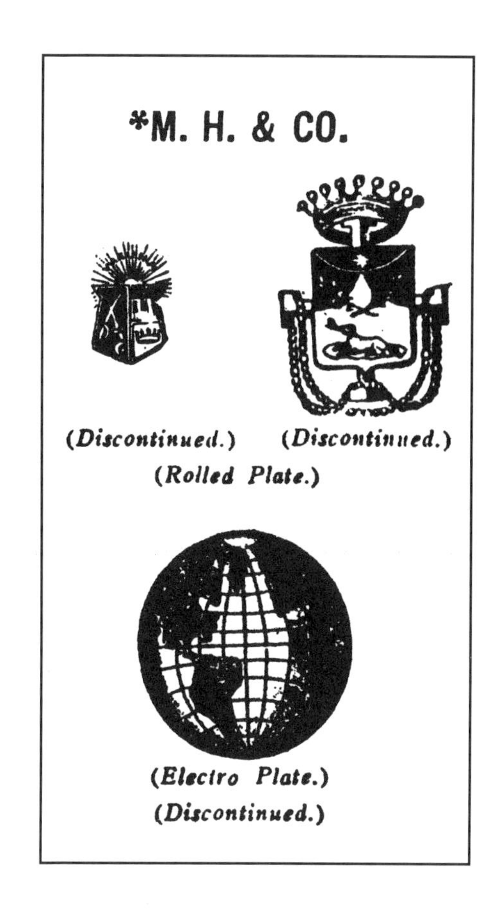

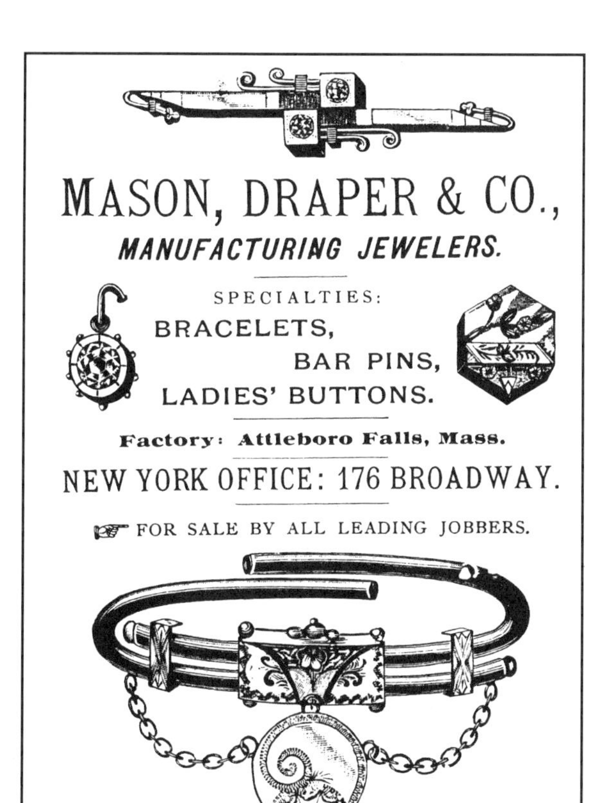

Mason, Draper & Co.
Attleboro Falls, Mass.
Bracelets, bar pins, ladies' buttons

THE JOBBERS' HANDBOOK. 85

MASON, HOWARD & CO.

Manufacturers of

Adjustable Bracelets

Also the Patented

Velvet Bracelet

BRACELETS, BUTTONS
FOBS, ETC.

ATTLEBORO, - MASS.

NEW YORK OFFICE: 180 Broadway.

Mason, Howard & Company
(JC 1904 & 1915)
Attleboro, Mass.
An advertisement of 1910 reads "Velvet bracelets, buttons, and fobs." The Velvet adjustable bracelet
was patented December 13, 1904.

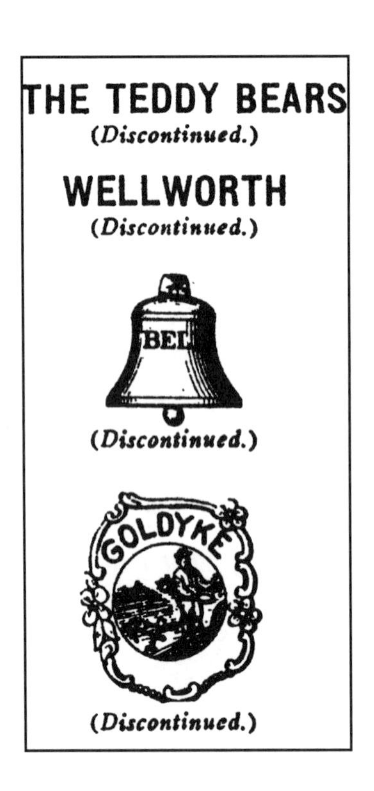

5058

Gold plated Lever Buttons. These are made of red and green gold, finely hand chased, set with brilliants. Sample pair by mail, 36 cents; one dozen pairs by express, $3.38.

5059

Gold plated Lever Buttons, embossed design, finished all bright. Sample pair by mail, 30 cents; one dozen pairs by express, $2.50.

5060

Gold plated Lever Buttons. One side plain, the other set with either moonstone, sapphire, ruby, amethyst, turquoise or topaz. Sample pair by mail, 30 cents; one dozen pairs by express, $2.50.

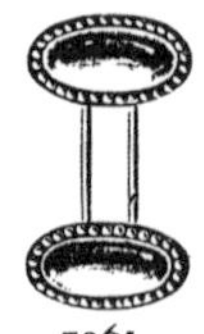

5061

Gold plated Link Buttons. Set on both sides with same assortment of stones as 5060. Sample pair by mail, 30 cents; one dozen pairs by express. $2.50.

5062

Gold plated Link Buttons. One side plain, the other set with either emerald, ruby, topaz or amethyst. Sample pair by mail, 30 cents; one dozen pairs by express, $2.50.

5063

Pearl Link Buttons. These are made from fine stock. Sample pair by mail, 22 cents; one dozen pairs by express, $2.00.

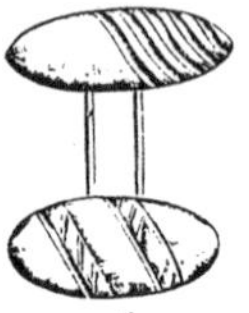

5064

Pearl Link Buttons. Best grade, finely carved. Sample pair by mail, 22 cents; one dozen pairs by express, $2.00.

5065

Gold filled Lever Sleeve Buttons. Heavily embossed, finely chased. These are rare value. Sample pair by mail, 70 cents; one dozen pairs by express. $6.75.

5066

Gold filled Lever Sleeve Buttons. Same quality and finish as 5065. Sample pair by mail, 70 cents; one dozen pairs by express, $6.75.

5067

Gold plated Lever Sleeve Buttons. Set with brilliants, hand chased, and finished bright. Sample pair by mail, 40 cents; one dozen pairs by express, $3.50.

5068

Gold plated Sleeve Buttons. Lever back, embossed, hand chased and finished bright. Sample pair by mail, 40 cents; one dozen pairs by express, $3.50.

ADDRESS McRAE & KEELER, ATTLEBORO, MASS.

New Goods in Bicycle Jewelry.

Realizing that we are in the midst of the Bicycle Age, we have turned our attention to the wants of Bicycle Riders, and as a result we are able to present a number of catchy novelties in this line at very Low Prices. All illustrations are exact size.

Bicycle Watch Charm.

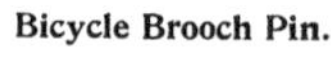

Bicycle Brooch Pin.

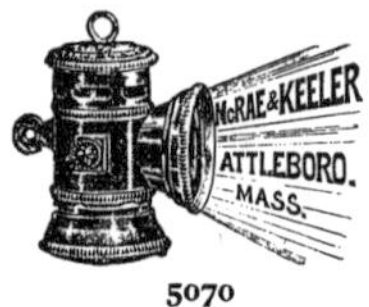

5070

We make this article with a ring in the top to hang on the watch chain, also with a clasp pin on the back; thus they can be worn on the waist, cap or coat. Please mention which is wanted when ordering. Sample, price each, silver or gold plate, 25 cents; per dozen $2.00.

5071

L. A. W. Buttons enamelled in White. Blue or Black. These buttons are screw backs, with spur to hold them in position. They are Gold Plated and finely finished. Usual price, 50 cents each. Our price, sample, 20 cents; per dozen, $1.50.

5072

This combination of bicycle and heart is very appropriate for lady riders. A portion of the bicycle frame is enamelled in black. We make them in Silver or Gold Plate. Sample, price each, 15 cents; per dozen, $1.20.

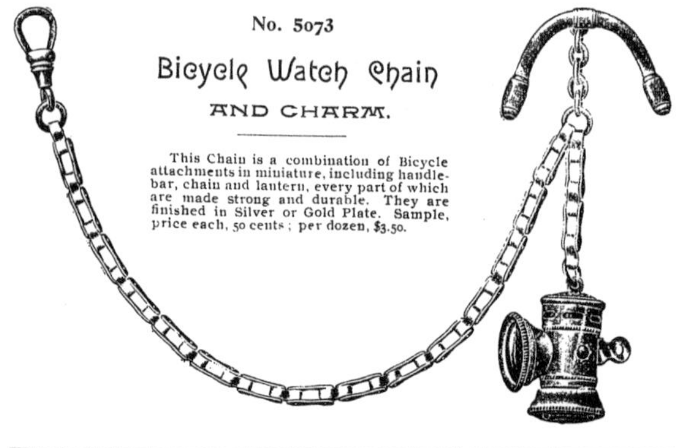

No. 5073

Bicycle Watch Chain

AND CHARM.

This Chain is a combination of Bicycle attachments in miniature, including handlebar, chain and lantern, every part of which are made strong and durable. They are finished in Silver or Gold Plate. Sample, price each, 50 cents; per dozen, $3.50.

ADDRESS McRAE & KEELER, ATTLEBORO, MASS.

23

McRae & Keeler
(c. 1895-, JC 1904-1915, K 1922 & 1931)
Attleboro, Mass.
Advertised to be the only Attleboro manufacturing jeweler with a catalog in 1897. Gold, sterling, and plated jewelry and novelties.

(1/20 *Gold Filled.*)

H.D. Merritt & Company
(c. 1855-1922)
North Attleboro, Mass.
Gold-filled chains.

Meyers, McNary & Company
("Out of business" before 1903)
Attleboro, Mass.
Gold plated chains.

Miller, Fuller, & Whiting
(JC 1909, JC 1915 "Out of business")
North Attleboro, Mass.

M. B.

Moore Bros.
(c. 1907-1940)
Attleboro, Mass.
Manufacturing jewelers.

M. M. CO.

Moore Manufacturing Company
(JC 1915)
Attleboro, Mass.

Edward H. Morse & Company
(JC-K 1943)
Attleboro, Mass.

MACO

Morse Andrews Company
(JC-K 1943)
Attleboro, Mass.

Morton Manufacturing Company
(K 1922 G.E. Morton & Co., 1931 Morton Mfg. Co.)
Attleboro, Mass.
Plated brass jewelry and costume jewelry including necklaces, bracelets, and earrings.

THE JOBBERS' HANDBOOK. 123

THE LEOMINSTER FINE TOOL AND MACHINE WORKS

Kendall Lane
LEOMINSTER, MASS.
Makers of
FINE TOOLS
for Celluloid Comb and Jewelry Manufacturers

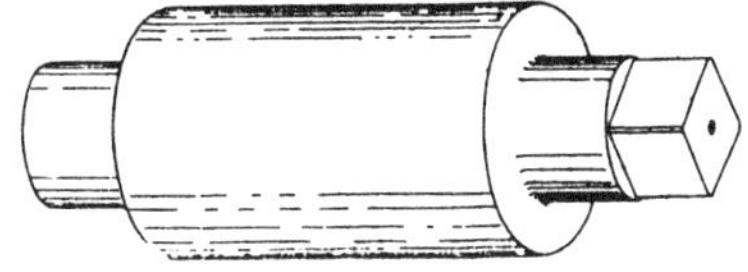

We are especially equipped to grind and lap jewelers' Rolls and Flattening Dies, producing the highest possible mirror finish. We make Jack Dies and Flat Dies to order.
QUICK DELIVERIES—LOW PRICES.
FRANK MOSSBERG CO.
ATTLEBORO, MASS.

Mossberg Manufacturing Company
(founded 1892)
Attleboro, Mass.
Providence, R.I.
Manufacturers of equipment used in jewelry manufacturing.

M.S. Company
(c. 1913-)
Attleboro, Mass.
Jewelry chain and findings.

NOVEL CRAFT

Novel Craft
(JC-K 1950)
Attleboro, Mass.

G. L. P. CO.

George L. Paine & Company
(JC 1909 & 1915, K 1922)
Solid gold, gold filled, and gold front jewelry. Belt buckles, scarf pins, and tie clips.

P. & P.
STERLING

Palmer & Peckham
(c. 1896-1935)
North Attleboro, Mass.

Paye & Baker Manufacturing Company
(c. 1898 succeeded Simmons & Paye)
North Attleboro, Mass.
Sterling souvenir spoons, flatware, and jewelry.
Sterling enamel. Match safes.

P Craft
Attleboro, Mass.
Costume jewelry.

L. S. P. CO.

LITTLE PRINCESS

L.S. Peterson Company, Inc.
(JC-K 1943-present)
North Attleboro, Mass.
Gold-filled "Little Princess" baby jewelry, rings, and silver novelties.

AMERICAN QUEEN

P. & K.

SECURITY

Pitman & Keeler
(K 1943 & 1950)
Attleboro, Mass.
Makers of the "American Queen" gold-filled expansion bracelet.

J. H. P.

J. H. PECKHAM & CO.
Makers of High Grade
Gold Filled Jewelry
in ROMAN and BRIGHT FINISH
84 Chestnut St., North Attleboro
New York: 9 Maiden Lane (Jewelers Bldg.)

J.H. Peckham & Co.
(c. 1910)
North Attleboro
An advertisement of 1910 reads "Makers of high grade gold-filled jewelry in roman and bright finish. Located at 84 Chestnut St., N. Attleboro, Office in New York."

P. S. CO.

Plainville Stock
Company
(c. 1872-present)
Plainville, Mass.
Gold, gold filled and
plated jewelry. Filigree
bracelets
and necklaces.

STERLING R. R. STERLING CRR

R.

卐 卐 STERLING CRR R.

(Discontinued.)

C. Ray Randall & Company
(JC 1909 & 1915, K 1922 & 1931, JC-K 1977-)
North Attleboro, Mass.
Solid gold, gold filled, and sterling jewelry. Bar pins, brooches, scarf pins, necklaces, belt buckles, bracelets.

Regnell, Bigney & Company
(c. 1894, "out of business" 1915)
Attleboro, Mass.
An advertisement of 1894 reads "makers of the leading fads of the day. Also a general line of ladies' goods, such as corsage pins, hair pins, hat pins, belt pins, belt buckles, side combs, tie pins, brooch & lace pins, etc.", illustrations of Palmer Cox "Brownie" pins.

Quaker Silver Company
(c. 1926-1959)
North Attleboro, Mass.
Sterling and silver plate
hollowware and novel-
ties.

Rhodes Bros. & Rothschild
(JC 1915 "out of business")
Attleboro, Mass.

SURE-LOCK

HOL-TITE

HANDY CLASP

AD-EX

La Mode

R & G Company (Ripley & Gowen)
(JC 1915, K 1922 & 1931)
Attleboro, Mass.
Jewelry, including filigree. Enamel vanity items-compacts, lipsticks, perfumes, with the trade name LaMode. Sterling enamel items.

(Discontinued.)

E. IRA R. & CO.,

E. Ira Richards & Company
(c. 1833 Ira Richards and Company, 1875 renamed E. Ira Richards & Company).
North Attleboro, Mass.
Jewelry manufacturing.

WRE

Symmetalic

W.E. Richards Company
(c. 1902-present)
North Attleboro, Mass.
Makers of "Symmetalic" jewelry combining sterling and gold.

R. MFG. CO.
LILLIPUTIAN

The Richards Manufacturing Company
(JC 1915)
Attleboro, Mass.

Stephen Richardson &
Company
(C. 1836-)
Attleboro, Mass.

R. & F.
THE VICTOR
(*On Buttons.*)

Riley & French
(JC 1904 Riley, French &
Heffron, c. 1909 became
Riley & French. Last
listing 1922.)
North Attleboro, Mass.

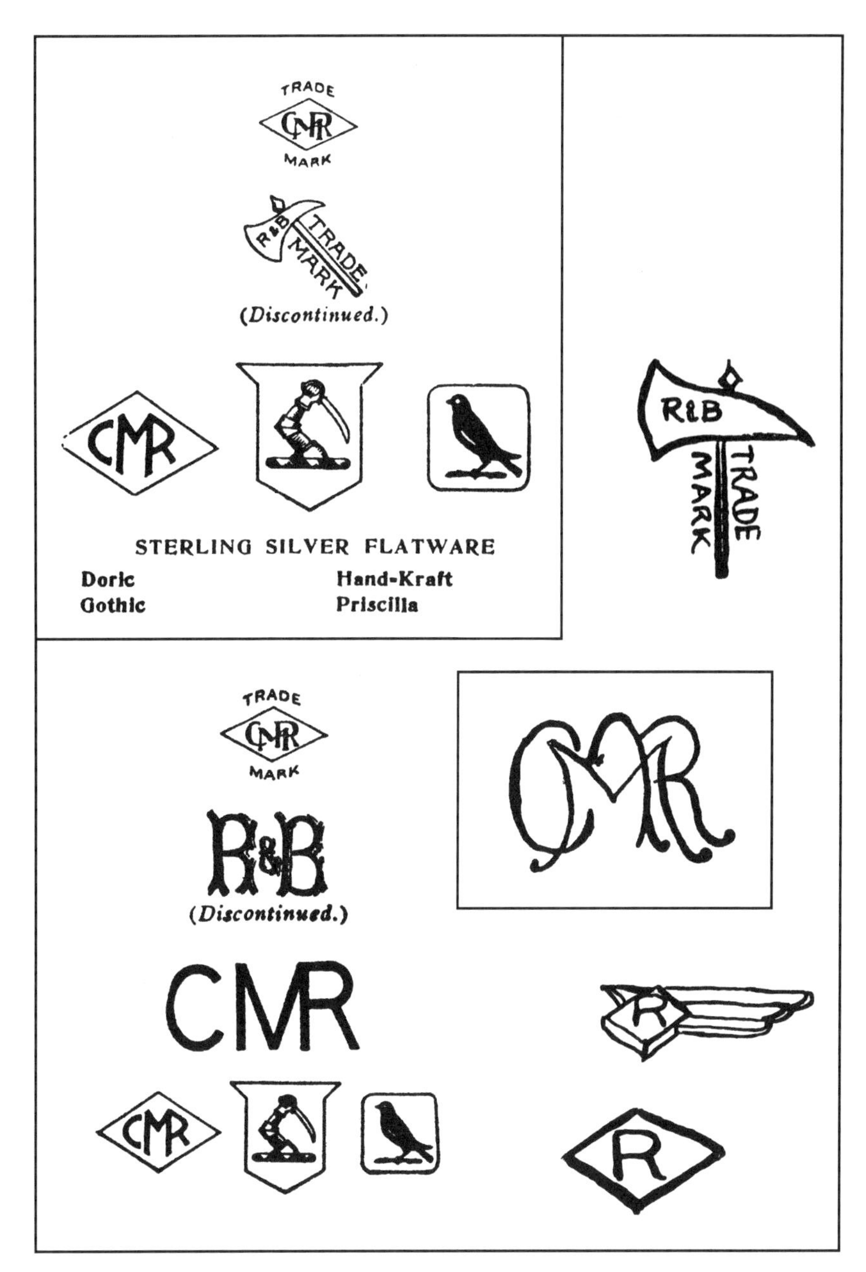

Charles M. Robbins Company
(c. 1892-present)
Attleboro, Mass.
Medals, emblems, awards. Sterling enamel jewelry. Souvenir spoons. At one time, their enameling department was the largest in the world.

Robinson & Company
(JC 1896 & 1904, JC
1915 "Out of business")
South Attleboro, Mass.

E. E. R. 10K
E. E. R. 14K

E.E. Rockwood
(K 1922 & 1931, JC-K 1943)
Attleboro Falls, Mass.
Watch and neck chains, fobs.

(Lockets.)

Rothschild Bros. & Company
(JC 1904, succeeded by Elias Rees & Son before 1915)
Attleboro, Mass.

STERLING SILVER TOILETWAR

Applied Bead	**Knickerbocker**
Belmont	**Plain**
Bronx	**Plymouth**
Cosmos	**Poppy**
Crescent	**Puritan**
Essex	**Rosaline**
Evangeline	**Rose**
Fairfax	**Sweet Pea**
Grape	**Victoria**
Grecian	**Virginia**

(Discontinued Patterns.)

Ideal	**Royal**
Queen	**Standard**

THE JOBBERS' HANDBOOK. 201

THE W. H. SAART CO.

SILVERSMITHS and MANUFACTURING JEWELERS

Toilet Goods, Buckles, Sash Pins, Coin Holders, Wrist Bags, Hat Pins, Match Boxes, Brooches, Beauty Pins, Manicure Goods, Bracelets, Novelties, Etc.

ATTLEBORO, MASS.

Salesrooms: New York, 49 Maiden Lane, Room 1002. Montreal: 204 St. James Street.

Saart Bros. Company
(c. 1905, JCK 1934)
Attleboro, Mass.
An advertisement of 1910 reads "Silversmiths and manufacturing jewelers--Toilet goods, buckles, sash pins, coin holders, wrist bags, hat pins, match boxes, brooches, beauty pins, manicure goods, bracelets, novelties, etc., Offices in New York and Montreal."

WILLIAM BENS CO.

.. Silversmiths ..

STERLING SILVER TOILET WARES and MANICURE GOODS

95 Chestnut Street,
PROVIDENCE, RHODE ISLAND

New York Office: 396 Broadway
Chicago Office: 501 Heyworth Bldg.
San Francisco Office: Jewelers' Bldg.

(Discontinued.)

***BORNEO**

(Discontinued.)

(Rings.)

THE JOBBERS' HANDBOOK. 263

F. H. SADLER CO.

Manufacturers of

THE . . .  BRACELETS

BANNER BRACELET

Lowest Priced Adjustable Bracelet on the Market.

Also Makers of High Grade Gold Filled

Scarf Pins, Scarf and Link Sets, Sash Pins, Hat Pins, Etc.

Factory: ATTLEBORO, MASS.

NEW YORK OFFICE: 180 Broadway, Room 43.

F.H. Sadler
(JC 1896-1915, K 1922)
Attleboro, Mass.
An advertisement of 1910 reads "Manufacturers of the Norma bracelet-Banner bracelet-lowest priced adjustable bracelet on the market. Also makers of high grade gold filled. Scarf pins, scarf and link sets, sash pins, hat pins, etc."

(Combs.)

Sadler Bros., Inc.
(c. 1863-present)
Attleboro, Mass.
Plated jewelry, hair ornaments, chains, pendants and bag frames.

S. M. & S.
(On Cards Only.)

Scofield, Melcher, & Scofield
(JC 1904 & 1915, K 1922, successors to Lincoln, Bacon & Co. before 1915)
Plainville, Mass.
Gold jewelry.

E.A. Scott
(K 1931)
Attleboro, Mass.
Jewelry.

SHEFFIELD

Sheffield, Inc.
(JC-K 1950)
Attleboro, Mass.

F. L. S. & CO.

F.L. Shephardson & Company
(JC 1904 & 1915)
North Attleboro, Mass.
Plated chains and jewelry.

Shields, Inc.
(c. 1920 as Fillkwik Company, 1939 purchased by Rex Products Corp. of New Rochelle, N.Y., 1957 they purchased Volupte of New Jersey)
Insignia and medals during World War II, later men's jewelry.

ARMILLA
BETSY ROSS
BRENDA
FLORADORA
SLIDENT
STUBBY
TYTON
VENETIAN
VICTORIAN

R.F. Simmons Company
(c. 1873-present)
Attleboro, Mass.
Watch chains, chateleine pins, eye-glass chains, fobs, and bracelets.

(Discontinued.)

ARMILLA
(Bracelets.)

R. F. S. & CO.
(Discontinued.)

R. F. S. CO.
(Discontinued.)

"SIMMONS' CHAINS"
(Discontinued.)

***SIMMONS**
TRADE MARK
CHAINS

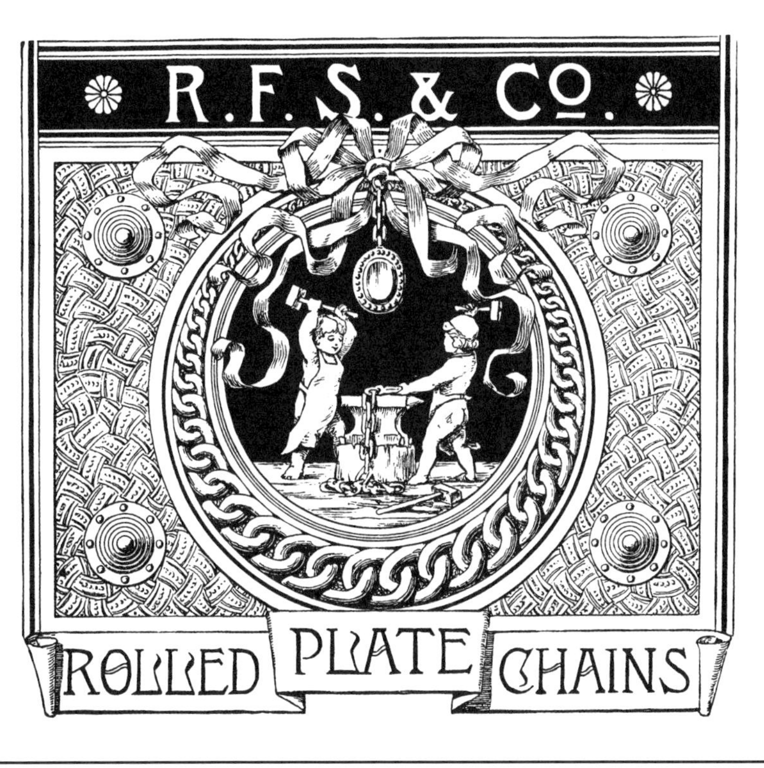

B. S. O. CO.

Short, Nerney & Company
(JC 1896)
Attleboro, Mass.

S & P

Simmons & Paye
(c. 1896, succeeded by Paye & Baker)
North Attleboro, Mass.

E. A. S. & CO.

E.A. Slade & Company
(c. 1915)
Attleboro, Mass.
Jewelry.

(Discontinued.)

T.I. Smith & Company
(JC 1904-1915, K 1922)
North Attleboro, Mass.

J. J. S. (on swivels)   **J. J. S. & CO.** (on swivels)

J. J. S.
J. J. S. & CO.
* 幸福
(Discontinued.)

(Discontinued.)

S. M. CO.
(Discontinued.)

S. & M.
(Discontinued.)

S. & M. CO.
(Discontinued.)

Sommer & Mills Company
(succeeded by J.J. Sommer & Company
before 1904)
North Attleboro, Mass.

S. & C.

Smith & Crosby
(JC 1904-1915, K 1922
&1931)
Attleboro, Mass.
Gold filled jewelry
including cufflinks, scarf
pins, and brooches.

D. S. S.
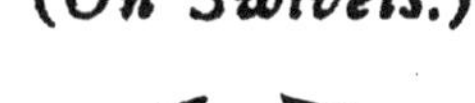
(On Swivels.)

D.S. Spaulding
(JC 1904 & 1915)
Mansfield, Mass.
(JC 1904 & 1915)
Plated chains and
swivels.

Standard Button Company
(JC 1909 & 1915, K 1922)
Attleboro, Mass.
Plated jewelry.

S J Co
(on jewelry)

Standard Jewelry
Company
(K 1931, JC-K 1943)
Attleboro, Mass.
Gold and gold-filled
jewelry.

THE JOBBERS' HANDBOOK. 125

SMITH & RICHARDSON
Manufacturers of
MACHINE AND WOVEN WIRE
.. CHAINS ..
ALSO SAFETY FOB FASTENERS AND WOVEN WIRE BRACELETS
Also Manufacturers of the
IMPERIAL POLISHING TANK, which Burnishes and Polishes JEWELRY, SILVERWARE, OPTICAL GOODS and NOVELTIES
IT LEADS THEM ALL
Send for Descriptive Circular
36 Railroad Street, ATTLEBORO, MASS.
Manufacturers should see the line

Smith & Richardson
Attleboro, Mass.
An advertisement of 1910 reads "Manufacturers of machine and woven wire chains-also safety fob fasteners and woven wire bracelets. Also manufacturers of the imperial polishing tank, which burnishes and polishes jewelry, silverware, optical goods and novelties-it leads them all."

Standard Chain Company
(c. 1923-)
North Attleboro, Mass.
Milanese mesh, gold,
sterling, and costume.

AGACITE
STAN-GLO

Standard Plastics
Company
(JC-K 1950-)
Attleboro, Mass.
Imitation stones.

S. BROS.
(On Tags.)

Stanley Bros.
(Out of business before 1904)
Attleboro Falls, Mass.
Plated chains.

S. & W.

Sturtevant & Whiting
Company
(descended from S.E.
Fisher & Company,
founded 1867)
North Attleboro, Mass.
Rolled gold plate jewelry,
including necklaces,
pins, and link buttons.

Howard H. Sweet & Son
(c. 1953-)
Attleboro, Mass.
Jewelry chain from
precious metals.

-S-

F.B.S.

-S-

F.B.S.

F.B. Stanton Company
(JC 1915)
Attleboro, Mass.
Jewelry.

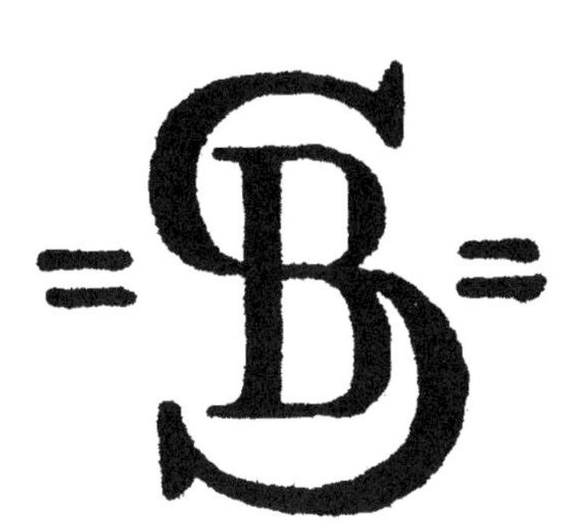

Streeter Brothers
(JC 1896, JC 1904 "Out
of business")
Attleboro, Mass.
Jewelry

'ARISTO-GRAM
ELBO-LINK
HOL-TITE
KUM-A-PART
LOOP LINKS
NU-LOK
PIN KLIP
RO-LON
BAER & WILDE

Swank, Inc.
(1897 Attleboro Mfg. Co., 1908 Baer & Wilde Company,
1931-present Swank).
Attleboro, Mass.
Primarily known for men's jewelry and accessories. The
Kum-a-part cuff button was first manufactured in 1918 and
enjoyed great success for many years.

J. F. S. S.
(On Swivels.)
(Discontinued.)

(On Carded Goods.)

(The following have been discontinued.)
THE DEFENDER
FAIR HARVARD
HERE'S A GOOD OLD YALE
LITTLE BILLIE
PRINCETON

(Discontinued.)

J.F. Sturdy's Sons
Company
(c. 1849- 1950)
Attleboro Falls, Mass.
Jewelry, especially
chains.

TRADE
MG
MARK

Sweet Manufacturing Company
(c. 1898-)
West Mansfield, Mass.
Frank Sweet developed an automatic chain-making machine, which made better chain, at a faster speed, than other equipment of the time, c. 1900. He established the Electric Chain Company in Attleboro, which was sold within four years, and the Sweet Manufacturing Company was organized.

THE JOBBERS' HANDBOOK. 89

C. O. SWEET & SON CO.
Manufacturing Jewelers
ATTLEBORO, MASS.
HIGH GRADE JEWELRY AND NOVELTIES
For Ladies' Wear
GEO. A. SCHAEFER, 9-11 Maiden Lane, New York.
F. D. WHITE, 704 Heyworth Bldg., Chicago, Ill.
FRED L. LEZINSKY CO., 704 Market St., San Francisco.

THOMAS BREESE
ENGRAVER, ENAMELER
ENAMEL PAINTER on Fine Jewelry
Emblems - College Flags - Specialties
Manufacturer and Importer of
Fine Enamel for Gold, Silver and Brass
AT LOW PRICES
322 Park St. ATTLEBORO, MASS.
Telephone

C.O. Sweet & Son Company
(JC 1909 & 1915, K 1922 & 1931)
An advertisement of 1910 reads "Manufacturing jewelers of high grade jewelry and novelties for ladies' wear."

Swift & Fisher
North Attleboro, Mass.
(JC 1915, K 1922 & 1931, JC-K 1943 & 1965)
Jewelry, rosaries, watch bracelets.

Sykes & Strandberg
(JC 1909 & 1915, K 1922 & 1931)
Attleboro, Mass.
Gold front jewelry.

Alexander F. Tanner
(K 1931)
Attleboro, Mass.
Sterling costume jewelry.

STERLING SILVER TOILETWARE
600 Windsor
800 Chippendale
900 Canterbury
000 Yorkshire
100 Intaglio
200 Martha Washington
300 The Warwick
400 Antique

(Discontinued Patterns.)
00 Meadow Rose
10 Rose of Sharon

TRADE

STERLING

MARK

Charles Thomae & Son, Inc.
(c. 1920-present)
Attleboro, Mass.
Charles Thomae was superintendant for the Watson Company, where he developed the enamel process for jewelry, novelties, and dresserware with such success that the Thomae Company was formed as a division of the Watson Company. Charles Thomae resigned from both companies and opened Charles Thomae & Son with his son, Herbert L. Thomae in 1920. Makers of sterling and gold accessories. Pins, cufflinks, money clips, pill boxes, hand mirrors, cufflink and cigar boxes.

Tifft & Whiting
(c. 1840, succeeded by Whiting Manufacturing Co., Newark, New Jersey in 1866, purchased by Gorham in 1926)
North Attleboro, Mass.

THE JOBBERS' HANDBOOK. 331

"GROBET SWISS FILES"
and other High Grade Tools for JEWELERS and SILVERSMITHS
Importers of Walrus Leather
MONTGOMERY & CO.
105-107 Fulton St., New York City.

F. L. TORREY & CO.
Manufacturers of
GENERAL LINE OF
Ladies' and Gents' Jewelry
Carefully designed and finished. Artistically Engraved. A line that wears well and looks good
FOR JOBBERS ONLY
ROBINSON BUILDING No. 2
ATTLEBORO, -:- MASS.

F.L. Torrey
(c.1910)
An advertisement from 1910 reads "Manufacturing a general line of ladies and gents jewelry, carefully designed and finished. Artistically engraved. A line that wears well and looks good. Robinson Building #2". Many Attleboro jewelry makers rented space in factories owned by their competitors.

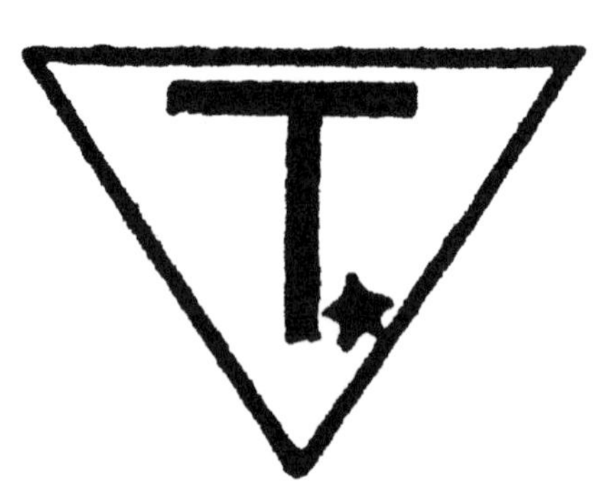

Totten Manufacturing Company
(JC 1915)
North Attleboro, Mass.
Jewelry.

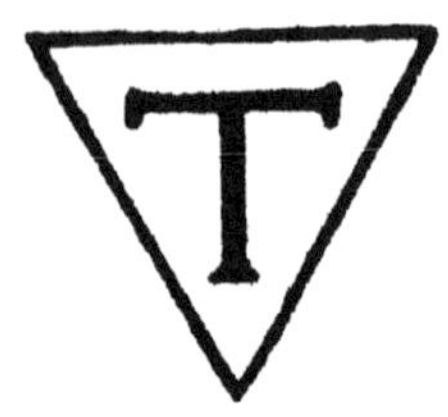

(Discontinued Jan. 1, 1897:)

Totten & Sommer Company
(JC 1896, succeeded by J.J. Sommer in 1897).
North Attleboro, Mass.

Tyndall Bros.
(c. 1910)
Attleboro, Mass.
An advertisement of 1910 reads "A popular priced line of sash pins, veil pins, handy pins, pendants, dutch collar pins, hat pins, scarf pins, brooches, and mesh bags. Also a rolled plate line of jewelry."

Unity Manufacturing
Company
(K 1931)
Attleboro, Mass.
Manufacturers of novelty
jewelry.

W.& R. Jewelry Company
(c. 1931)
Attleboro, Mass.
Jewelry manufacturers.

(Discontinued.)

The White Manufacturing Company
(K. 1931 & JC-K 1943 & 1950)
North Attleboro, Mass.
Rolled gold plate and
sterling silver jewelry.

G.K. Webster
(c. 1879, 1892 succeeded by the Webster
Company)
North Attleboro, Mass.
Sterling silver novelties and dresserware.

TOP HAT

Wells, Inc.
(c. 1922, c. 1967 purchased R. Blackinton &
Company, c. 1977 combined with Benrus to form
Wells Benrus Corp., left jewelry industry 1978)
Attleboro, Mass.
Sterling charms and other jewelry. Advertised as the
world's largest manufacturer of moveable charms.

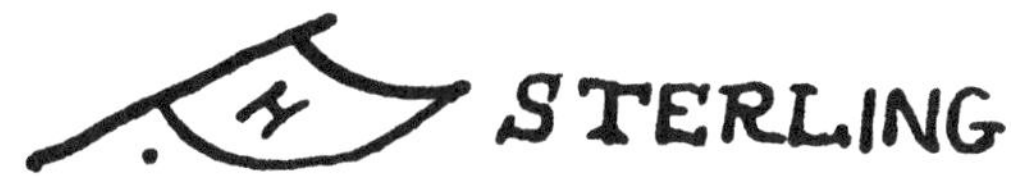

Watson Company
(c. 1874 Cobb & Gould, 1894-1910 partnership of C.L.
Watson and Fred A. Newell as Watson & Newell Co.,
Watson Company through 1955).
Attleboro, Mass.
Sterling silver souvenir spoons and enamel jewelry.
Flatware and hollowware.

H. Wexel & Company
(JC 1896, JC 1904 "Out
of business")
Buttons, studs, and
other jewelry.

F.W. Weaver & Company
(c. 1915)
Attleboro, Mass.
Jewelry.

W. & R.

White & Rounsville
(JC. 1915)
Attleboro, Mass.
Jewelry.

W. C. CO.

Whiting Chain Company
(JC 1915, K 1922)
Plainville, Mass.
Chains and lavalieres.

W. & D.
(Discontinued.)

ALICE NIELSEN
(Discontinued.)

FLORODORA
(Discontinued.)

YO SAN
(Discontinued.)

Whiting & Davis Company
(c. 1876-present)
Plainville, Mass.
In 1876, the company began as a small chain manufacturer. For many years Whiting & Davis crafted handmade mesh until the company developed the first chainmail mesh machine in 1907 and later grew to become the world's largest manufacturer of mesh products. Their home office is located in N. Attleboro today.

(Present Mark.)

(Discontinued.)

STERLING SILVER FLATWARE

Bead Unique	Marlborough
Bird	Marquis
Bow Knot	Narcissus
Damascus	Neapolitan
Ester	Orleans
Flemish	Oxalis
Florence	Palm
Geneva	Plain Tipped
George III.	Portia
Gothic	Priscilla
Grape Vine	Roderick
Gladstone	Syria
Hagle	Tyrolean
Helena	Unique
La Fayette	Unique Bead
Lily of the Valley	Wheat

(Discontinued Flatware.)

Antique	Pointed Antique
Floral	Puritan
Josephine	Rose
Lily	

STERLING SILVER TOILETWARE

Athene	La Fayette
Colonial	Priscilla
Duchess	Queen Anne
Floral	Virginia

FRANK M. WHITING & CO.,
NORTH ATTLEBORO, MASS.

Frank M. Whiting & Company
(JC 1915)
North Attleboro, Mass.
Makers of sterling silver flatware and jewelry in the medallion pattern and figural designs.

Wilcox & Wagoner Co.
(JC 1915, Watson company successors)
Attleboro, Mass.

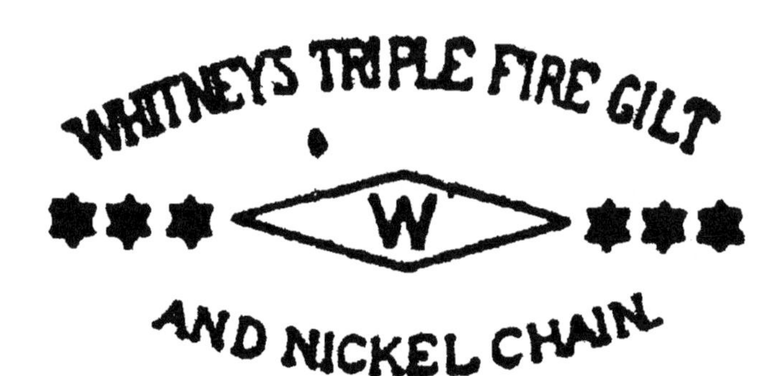

F.G. Whitney & Company
(JC 1896, JC 1904 "Out of business")
Attleboro, Mass.
Chain manufacturers.

Wearite

Winnstock Company
(K 1922)
Attleboro, Mass.
Jewelry.

WINTHROP MFG. CO.
Manufacturing Jewelers
Makers of
Artistic Goods in Original Designs
with the finest possible finish
FOBS, SASH PINS,
BROOCHES, HAT PINS,
SCARF PINS, PENDANTS
ETC.
For the Jobbing Trade Only.
ATTLEBORO, - - MASS.

Winthrop Manufacturing Company
(JC 1915)
Attleboro, Mass.
An advertisement of 1910 reads "Artistic goods in original designs with the finest possible finish-fobs, sash pins, brooches, hat pins, scarf pins, pendants."

THE JOBBERS' HANDBOOK. 101

W. H. WILMARTH & CO.
CORPORATION

Lever Links;
Separable, Lever and Automatic
Sleeve Buttons;
Lever, Standard and Separable
Collar Buttons;
Ladies' and Gents' Sets;
Full Line Ladies' and Gents'
Vest and Fob Chains;

FOR JOBBING TRADE ONLY

COR. SCHOOL AND HAZEL STREETS
ATTLEBORO, MASS.
New York Office: 180 Broadway

W.H. Wilmarth & Company
(JC 1897-1915, K 1931)
Attleboro, Mass.
An advertisement of 1910 reads " Lever links; separable, lever, and automatic sleeve buttons; lever, standard, and separable collar buttons; ladies and gents' sets: full line ladies and gents' vest and fob chains. Office in New York."

P. E. WITHERELL
ATTLEBORO, MASS.
MANUFACTURER OF
150 Different Patterns
OF

442 439
Rolled Gold Plate Collar Buttons
WITH ENAMEL BACKS
These are High Grade Buttons and Warranted
Not to Soil the Linen or Poison the Skin.

P.E. Witherell
(JC 1904 & 1915)
Attleboro, Mass.
An advertisement of 1910 reads " Manufacturers of 150 different patterns of rolled gold plate collar buttons with enamel backs."

Young, Bagnall & Company
(c. 1870, succeeded by Doran, Bagnall & Company before 1905)
North Attleboro, Mass.
Manufacturing jewelers.

Bibliography

Clayton, Larry. *The Evans Book*. Atglen, Pennsylvania: Schiffer Publishing Ltd., 1998.

Daggett, John. *A Sketch of the History of Attleborough from Its Settlement to the Division.* Boston, Massachusetts: Press of Samuel Usher, 1894.

Kovel, Ralph and Terry. *Kovels' American Silver Marks.* New York: Crown Publishers, Inc., 1989.

Lanpher, Bob, Dorothea Donnelly, and George Cunningham. *Images of America-North Attleborough, Massachusetts.* Dover, New Hampshire: Arcadia Publishing, 1998.

Rainwater, Dorothy. *American Jewelry Manufacturers.* Atglen, Pennsylvania: Schiffer Publishing Ltd., 1988.

Rainwater, Dorothy. *American Silver Manufacturers.* Atglen, Pennsylvania: Schiffer Publishing Ltd., 2001

Stone, Orra L. *History of Massachusetts Industries-Their Inception, Growth, and Success-Volume 1.* Boston Massachusetts and Chicago, Illinois: S. J. Clarke, 1930

Studley, A. Irvin. *Studley's History of Attleboro.* Attleboro, Massachusetts: Attleboro Historical Commission and Attleboro Historical Volunteers, 1994.

Tedesco, Dr. Paul H. *1894 Attleborough-Attleboro 1978-Hub of the Jewelry World.* Danvers, Massachusetts: Bradford & Bigelow, Inc., 1979.

Additional Reading for Buttons

Albert, Alphaeus. *Record of American Uniform and Historical Buttons.* Boyertown, Pennsylvania: Boyertown Publishing Co., 1976.

Fennimore, Donald. *Metalwork in Early America.* Winterthur, Delaware: Winterthur Publications, 1996.

Hughes, Elizabeth and Marion Lester. *The Big Book of Buttons.* Boyertown, Pennsylvania: Boyertown Publishing Co., 1981.

McGuinn, William and Bruce Bazelon. *American Military Button Makers and Dealers; Their Backmarks & Dates.* Chelsea, Michigan: BookCrafters, Inc., 1992.

Rulau, Russell. *Standard Catalog of U.S. Tokens 1700-1900.* Iola, WI: Krause Publications Inc., 1994.

Tice, Warren. *Uniform Buttons of the United States 1776-1865.* Gettysburg, Pennsylvania: Thomas Publications, 1997.

Thomas' Register of American Manufacturers and First Hands in All Lines. New York, N.Y.: Thomas Publishing Co., 1905.

Further Reading and Other Resources for Match Safes

International Match Safe Association
P. O. Box 791
Malaga, New Jersey 08328
Phone: 856-694-4167
E-mail: IMSAoc@aol.com
Website: www.matchsafe.org

IMSA is open to anyone with an interest in match safes. Worldwide membership ranges from novice to some of the world's premiere collectors, dealers, authors and specialists. Quarterly newsletter, annual convention, commemorative match safe and worldwide networking opportunities.

Alsford, Denis. *Match Holders 100 Years if Ingenuity.* Atglen, Pennsylvania: Schiffer Publishing Ltd., 1994.

Crosby, Deborah. *Silver Novelties of the Gilded Age: 1870-1910.* Atglen, Pennsylvania: Schiffer Publishing Ltd., 2001.

Fresco-Corbu, Roger. *Vesta Boxes.* Cambridge, England: Lutterworth Press, 1983.

Patterson, J. E. *Matchsafes in the Collection of the Cooper Hewitt Museum.* New York, New York: Eastern Press, 1981.

Sanders, Jr., W. Eugene and Christine Sanders. *Pocket Matchsafes, Reflections of Life and Art 1840-1920.* Atglen, Pennsylvania: Schiffer Publishing Ltd., 1997.

Shinn, Deborah Sampson. *Matchsafes – Smithsonian Cooper-Hewitt, National Design Museum,* London, England: Scala Publishers Ltd., 2001.

Sullivan, Audrey G. *A History of Match Safes in the United States.* Fort Lauderdale, Florida: Riverside press, Inc., 1978.